# COMPONENT FAILURES, MAINTENANCE AND REPAIR

# COMPONENT FAILURES, MAINTENANCE AND REPAIR

Edited by

## M. J. NEALE
OBE, BSc(Eng), DIC, FCGI, WhSch, FEng, FIMechE

## A TRIBOLOGY HANDBOOK

Butterworth-Heinemann Ltd
Linacre House, Jordan Hill, Oxford OX2 8DP

A member of the Reed Elsevier plc group

OXFORD LONDON BOSTON
MUNICH NEW DELHI SINGAPORE SYDNEY
TOKYO TORONTO WELLINGTON

First published 1995
Reprinted 1995

**British Library Cataloguing in Publication Data**
Component Failures, Maintenance and
   Repair: Triology Handbook – (Triology
   Handbook)
   I. Neale, M.J.   II. Series
   621.8

ISBN 0 7506 0980 X

Printed in Great Britain

# Contents

# Editor's Preface

This handbook gives practical guidance on component failures, maintenance and repair in a form intended to provide easy and rapid reference. It is hoped that it will provide useful practical guidance for engineers concerned with plant maintenance and also design and development engineers in industry. It is based on material published in the first edition of the *Tribology Handbook* and has been updated and matched to international requirements.

Each section has been written by an author who is expert in the field and who in addition to understanding the related basic principles, also has extensive practical experience in his subject area.

The individual contributors are listed and the editor gratefully acknowledges their assistance and that of all other people who have helped him in the checking and compilation of this revised volume.

Michael Neale
Neale Consulting Engineers Ltd
Farnham 1994

# Contributors

| Section | Author |
|---|---|
| Failure patterns and failure analysis | J. D. Summers-Smith BSc, PhD, CEng, FIMechE<br>M.J. Neale OBE, BSc(Eng), DIC, FCGI, WhSch, FEng, FIMechE |
| Plain bearing failures | P. T. Holingan BSc(Tech), FIM |
| Rolling bearing failures | W. J. J. Crump BSc, ACGI, FInstP |
| Gear failures | T. I. Fowle BSc(Hons), ACGI, CEng, FIMechE<br>H. J. Watson BSc(Eng), CEng, MIMechE |
| Piston and ring failures | M. J. Neale OBE, BSc(Eng), DIC, FCGI, WhSch, FEng, FIMechE |
| Seal failures | B. S. Nau BSc, PhD, ARCS, CEng, FIMechE, MemASME |
| Wire rope failures | S. Maw MA, CEng, MIMechE |
| Brake and clutch failures | T. P. Newcombe DSc, CEng, FIMechE, FInstP<br>R. T. Spurr BSc, PhD |
| Fretting problems | R. B. Waterhouse MA, PhD, FIM |
| Maintenance methods | M. J. Neale OBE, BSc(Eng), DIC, FCGI, WhSch, FEng, FIMechE |
| Condition monitoring | M. J. Neale OBE, BSc(Eng), DIC, FCGI, WhSch, FEng, FIMechE |
| Operating temperature limits | J. D. Summers-Smith BSc, PhD, CEng, FIMechE |
| Vibration analysis | M. J. Neale OBE, BSc(Eng), DIC, FCGI, WhSch, FEng, FIMechE |
| Wear debris analysis | M. H. Jones BSc(Hons), CEng, MIMechE, MInstNDT<br>M. J. Neale OBE, BSc(Eng), DIC, FCGI, WhSch, FEng, FIMechE |
| Performance analysis | M. J. Neale OBE, BSc(Eng), DIC, FCGI, WhSch, FEng, FIMechE |
| Allowable wear limits | H. H. Heath FIMechE |
| Repair of worn surfaces | G. R. Bell BSc, ARSM, CEng, FIM, FWeldI, FRIC |
| Wear resistant materials | H. Hocke CEng, MIMechE, FIPlantE, MIMH, FIL<br>M. Bartle CEng, MIM, DipIM, MIIM, AMWeldI |
| Repair of plain bearings | P. T. Holligan BSc(Tech), FIM |
| Repair of friction surfaces | T. P. Newcomb DSc, CEng, FIMechE, FInstP<br>R. T. Spurr BSc, PhD |
| Industrial flooring materials | A. H. Snow FCIS, MSAAT |

## THE SIGNIFICANCE OF FAILURE

Failure is only one of three ways in which engineering devices may reach the end of their useful life.

| The way in which the end of useful life is reached | | Typical devices which can end their life this way |
|---|---|---|
| Failure | Slow | Seals leaking progressively |
| | Sudden | Electric lamp bulbs |
| Obsolescence | | Gas lamps<br>Steam locomotives |
| Completion | | Bullets and bombs<br>Packaging |

In the design process an attempt is usually made to ensure that failure does not occur before a specified life has been reached, or before a life limit has been reached by obsolescence or completion. The occurrence of a failure, without loss of life, is not so much a disaster, as the ultimate result of a design compromise between perfection and economics.

When a limit to operation without failure is accepted, the choice of this limit depends on the availability required from the device.

*Availability* is the average percentage of the time that a device is available to give satisfactory performance during its required operating period. The availability of a device depends on its reliability and maintainability.

*Reliability* is the average time that devices of a particular design will operate without failure.

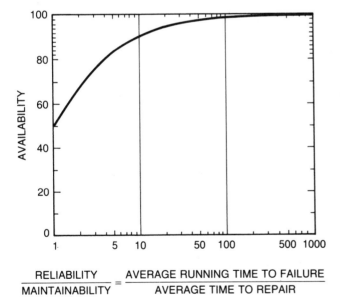

$$\frac{\text{RELIABILITY}}{\text{MAINTAINABILITY}} = \frac{\text{AVERAGE RUNNING TIME TO FAILURE}}{\text{AVERAGE TIME TO REPAIR}}$$

**Figure 1.1 The relationship between availability, reliability and maintainability.** *High availabilities can only be obtained by long lives or short repair times or both*

*Maintainability* is measured by the average time that devices of a particular design take to repair after a failure.

The availability required, is largely determined by the application and the capital cost.

## FAILURE ANALYSIS

The techniques to be applied to the analysis of the failures of tribological components depend on whether the failures are isolated events or repetitive incidents. Both require detailed examination to determine the primary cause, but, in the case of repeated failures, establishing the temporal pattern of failure can be a powerful additional tool.

### Investigating failures

When investigating failures it is worth remembering the following points:

(a) Most failures have several causes which combine together to give the observed result. A single cause failure is a very rare occurrence.

(b) In large machines tribological problems often arise because deflections increase with size, while in general oil film thicknesses do not.

(c) Temperature has a very major effect on the performance of tribological components both directly, and indirectly due to differential expansions and thermal distortions. It is therefore important to check:

    Temperatures
    Steady temperature gradients
    Temperature transients

### Causes of failure

To determine the most probable causes of failure of components, which exist either in small numbers, or involve mass produced items the following procedure may be helpful:

1. Examine the failed specimens using the following sections of this Handbook as guidance, in order to determine the probable mode of failure.

2. Collect data on the actual operating conditions and double check the information wherever possible.

3. Study the design, and where possible analyse its probable performance in terms of the operating conditions to see whether it is likely that it could fail by the mode which has been observed.

4. If this suggests that the component should have operated satisfactorily, examine the various operating conditions to see how much each needs to be changed to produce the observed failure. Investigate each operating condition in turn to see whether there are any factors previously neglected which could produce sufficient change to cause the failure.

# Failure patterns and failure analysis

## Repetitive failures

Two statistics are commonly used:-

1. MTBF (mean time between failures)

$$= \frac{L_1 + L_2 + ... + L_n}{n}$$

where $L_1$, $L_2$, etc., are the times to failure and $n$ the number of failures

2. $L_{10}$ Life, is the running time at which the number of failures from a sample population of components reaches 10%. (Other values can also be used, e.g. $L_1$ Life, viz the time to 1% failures, where extreme reliability is required.)

MTBF is of value in quantifying failure rates, particularly of machines involving more than one failing component. It is of most use in maintenance planning, costing and in assessing the effect of remedial measures.

$L_{10}$ Life is a more rigorous statistic that can only be applied to a statistically homogeneous population, i.e. nominally identical items subject to nominally identical operating conditions.

## Failure patterns

Repetitive failures can be divided by time to failure according to the familiar 'bath-tub' curve, comprising the three regions: early-life failures (infantile mortality), 'mid-life' (random) failures and 'wear-out'.

Early-life failures are normally caused by built-in defects, installation errors, incorrect materials, etc.

Mid-life failures are caused by random effects external to the component, e.g. operating changes, (overload) lightning strikes, etc.

Wear-out can be the result of mechanical wear, fatigue, corrosion, etc.

The ability to identify which of these effects is dominant in the failure pattern can provide an insight into the mechanism of failure.

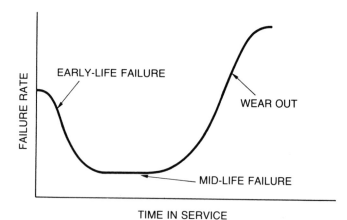

Figure 1.2 The failure rate with time of a group of similar components

As a guide to the general cause of failure it can be useful to plot failure rate against life to see whether the relationship is falling or rising.

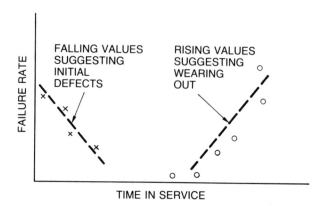

Figure 1.3 The failure rate with time used as an investigative method

## Weibull analysis

Weibull analysis is a more precise technique. Its power is such that it can provide useful guidance with as few as five repeat failures. The following form of the Weibull probability equation is useful in component failure analysis:

$$F(t) = 1 - \exp[\alpha(t - \gamma)^{\beta}]$$

where $F(t)$ is the cumulative percentage failure, $t$ the time to failure of individual items and the three constants are the scale parameter $(\alpha)$, the Weibull Index $(\beta)$ and the location parameter $(\gamma)$.

For components that do not have a shelf life, i.e. there is no deterioration before the component goes into service, $\gamma = 0$ and the expression simplifies to:

$$F(t) = 1 - \exp[\alpha t^{\beta}].$$

The value of the Weibull Index depends on the temporal pattern of failure, viz:

| | |
|---|---|
| early-life failures | $\beta = 0.5$ |
| random failures | $\beta = 1$ |
| wear out | $\beta = 3.4$ |

Weibull analysis can be carried out simply and quickly as follows:

1. Obtain the values of $F(t)$ for the sample size from Table 1.1
2. Plot the observed times to failure against the appropriate value of $F(t)$ on Weibull probability paper (Figure 1.5).
3. Draw best fit straight line through points.
4. Drop normal from 'Estimation Point' to the best fit straight line.
5. Read off $\beta$ value from intersection on scale.

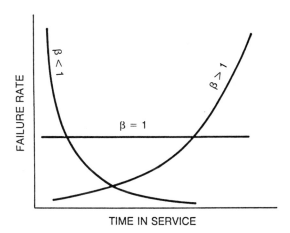

**Figure 1.4 The relationship between the value of $\beta$ and the shape of the failure rate curve**

For $n > 20$ – Calculate approximate values of $F(t)$ from

$$\frac{100(i - 0.3)}{n + 0.4}$$

where: $i$ is the $i$th measurement in a sample of $n$ arranged in increasing order.

Sample size

| | | | | | | | | | | | | | | | | | | | | |
|---|---|---|---|---|---|---|---|---|---|---|---|---|---|---|---|---|---|---|---|---|
| 5 | 12.9 | 31.3 | 50.0 | 68.8 | 87.0 | | | | | | | | | | | | | | | |
| 6 | 10.9 | 26.4 | 42.1 | 57.8 | 73.5 | 89.0 | | | | | | | | | | | | | | |
| 7 | 9.4 | 22.8 | 36.4 | 50.0 | 63.5 | 77.1 | 90.5 | | | | | | | | | | | | | |
| 8 | 8.3 | 20.1 | 32.0 | 44.0 | 55.9 | 67.9 | 79.8 | 91.7 | | | | | | | | | | | | |
| 9 | 7.4 | 17.9 | 28.6 | 39.3 | 50.0 | 60.6 | 71.3 | 82.0 | 92.5 | | | | | | | | | | | |
| 10 | 6.6 | 16.2 | 25.8 | 35.5 | 45.1 | 54.8 | 64.4 | 74.1 | 83.7 | 93.3 | | | | | | | | | | |
| 11 | 6.1 | 14.7 | 23.5 | 32.3 | 41.1 | 50.0 | 58.8 | 67.6 | 76.4 | 85.2 | 93.8 | | | | | | | | | |
| 12 | 5.6 | 13.5 | 21.6 | 29.7 | 37.8 | 45.9 | 54.0 | 62.1 | 70.2 | 78.3 | 86.4 | 94.3 | | | | | | | | |
| 13 | 5.1 | 12.5 | 20.0 | 27.5 | 35.0 | 42.5 | 50.0 | 57.4 | 64.9 | 72.4 | 79.9 | 87.4 | 94.8 | | | | | | | |
| 14 | 4.8 | 11.7 | 18.6 | 25.6 | 32.5 | 39.5 | 46.5 | 53.4 | 60.4 | 67.4 | 74.3 | 81.3 | 88.2 | 95.1 | | | | | | |
| 15 | 4.5 | 10.9 | 17.4 | 23.9 | 30.4 | 36.9 | 43.4 | 50.0 | 56.5 | 63.0 | 69.5 | 76.0 | 82.5 | 89.0 | 95.4 | | | | | |
| 16 | 4.2 | 10.2 | 16.3 | 22.4 | 28.5 | 34.7 | 40.8 | 46.9 | 53.0 | 59.1 | 65.2 | 71.4 | 77.5 | 83.6 | 89.7 | 95.7 | | | | |
| 17 | 3.9 | 9.6 | 15.4 | 21.1 | 26.9 | 32.7 | 38.4 | 44.2 | 50.0 | 55.7 | 61.5 | 67.2 | 73.0 | 78.8 | 84.5 | 90.3 | 96.0 | | | |
| 18 | 3.7 | 9.1 | 14.5 | 20.0 | 25.4 | 30.9 | 36.3 | 41.8 | 47.2 | 52.7 | 58.1 | 63.6 | 69.0 | 74.5 | 79.9 | 85.4 | 90.8 | 96.2 | | |
| 19 | 3.5 | 8.6 | 13.8 | 18.9 | 24.1 | 29.3 | 34.4 | 39.6 | 44.8 | 50.0 | 55.1 | 60.3 | 65.5 | 70.6 | 75.8 | 81.0 | 86.1 | 91.3 | 96.4 | |
| 20 | 3.4 | 8.2 | 13.1 | 18.0 | 22.9 | 27.8 | 32.7 | 37.7 | 42.6 | 47.5 | 52.4 | 57.3 | 62.2 | 67.2 | 72.1 | 77.0 | 81.9 | 86.8 | 91.7 | 96.5 |

**Table 1.1  Values of the cumulative percent failure F(t) for the individual failures in a range of sample sizes**

⊕ ESTIMATION POINT

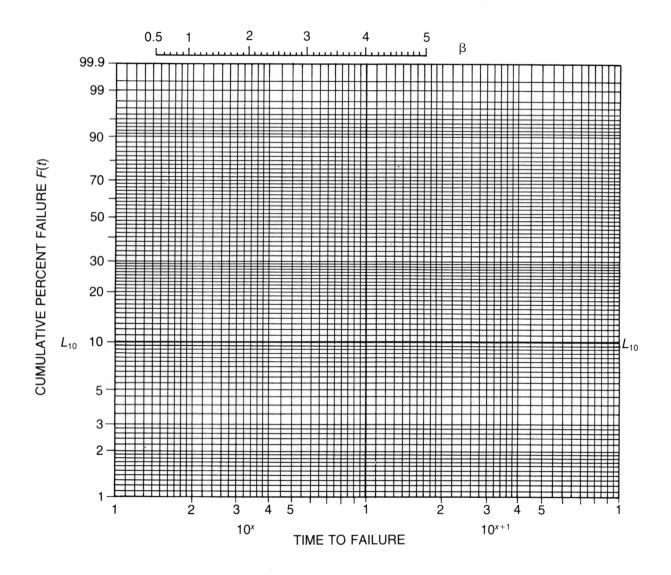

Figure 1.5 Weibull probability graph paper

Figure 1.6 gives an example of 9 failures of spherical roller bearings in an extruder gearbox. The $\beta$ value of 2.7 suggests wear-out (fatigue) failure. This was confirmed by examination of the failed components. The $L_{10}$ Life corresponds to a 10% cumulative failure. $L_{10}$ Life for rolling bearings operating at constant speed is given by:

$$L_{10} \text{ Life (hours)} = \frac{10^6}{n} \frac{C^x}{P}$$

Where $n =$ speed (rev/min), $C =$ bearing capacity, $P =$ equivalent radial load, $x = 3$ for ball bearings, 10/3 for roller bearings.

Determination of the $L_{10}$ Life from the Weibull analysis allows an estimate to be made of the actual load. This can be used to verify the design value. In this particular example, the exceptionally low value of $L_{10}$ Life (2500 hours) identified excessive load as the cause of failure.

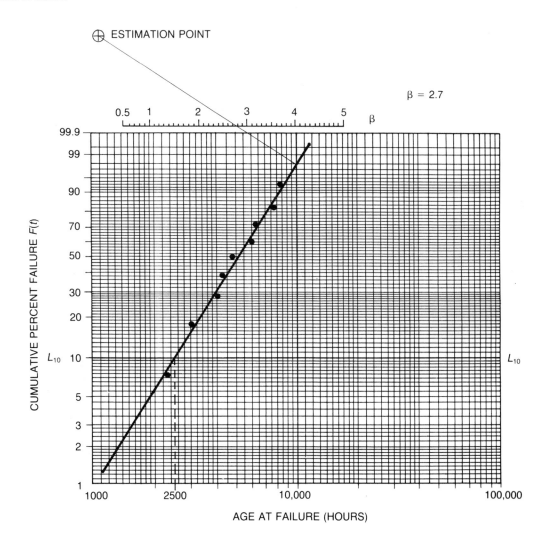

**Figure 1.6 Thrust rolling bearing failures on extruder gearboxes**

Figure 1.7 gives an example for 17 plain thrust bearing failures on three centrifugal air compressors. The $\beta$ value of 0.7 suggests a combination of early-life and random failures. Detailed examination of the failures showed that they were caused in part by assembly errors, in part of machine surges.

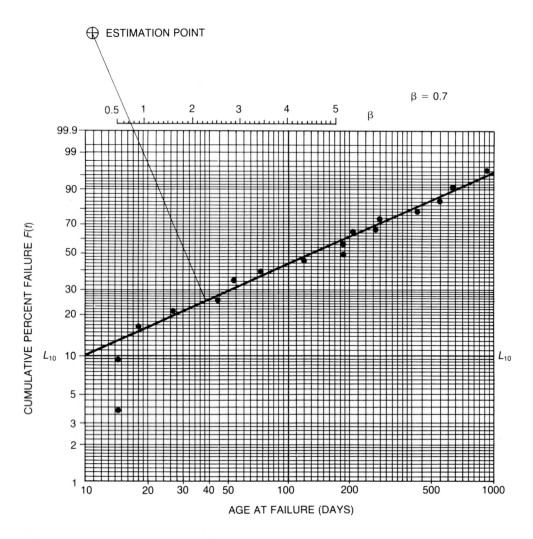

**Figure 1.7 Plain thrust bearing failures on centrifugal air compressors**

## Foreign matter

### Characteristics

Fine score marks or scratches in direction of motion, often with embedded particles and haloes.

### Causes

Dirt particles in lubricant exceeding the minimum oil film thickness.

## Wiping

### Characteristics

Surface melting and flow of bearing material, especially when of low-melting point, e.g. whitemetals, overlays.

### Causes

Inadequate clearance, overheating, insufficient oil supply, excessive load, or operation with a non-cylindrical journal.

## Foreign matter

### Characteristics

Severe scoring and erosion of bearing surface in the line of motion, or along lines of local oil flow.

### Causes

Contamination of lubricant by excessive amounts of dirt particularly non-metallic particles which can roll between the surfaces.

## Fatigue

### Characteristics

Cracking, often in mosaic pattern, and loss of areas of lining.

### Causes

Excessive dynamic loading or overheating causing reduction of fatigue strength; overspeeding causing imposition of excessive centrifugal loading.

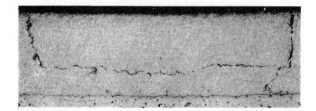

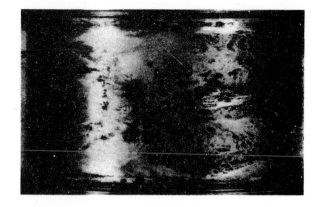

## Fatigue

### Characteristics

Loss of areas of lining by propagation of cracks initially at right angles to the bearing surface, and then progressing parallel to the surface, leading to isolation of pieces of the bearing material.

### Causes

Excessive dynamic loading which exceeds the fatigue strength at the operating temperature.

## Fretting

### Characteristics

Welding, or pick-up of metal from the housing on the back of bearing. Can also occur on the joint faces. Production and oxidation of fine wear debris, which in severe cases can give red staining.

### Causes

Inadequate interference fit; flimsy housing design; permitting small sliding movements between surfaces under operating loads.

## Excessive interference

### Characteristics

Distortion of bearing bore causing overheating and fatigue at the bearing joint faces.

### Causes

Excessive interference fit or stagger at joint faces during assembly.

## Misalignment

### Characteristics

Uneven wear of bearing surface, or fatigue in diagonally opposed areas in top and bottom halves.

### Causes

Misalignment of bearing housings on assembly, or journal deflection under load.

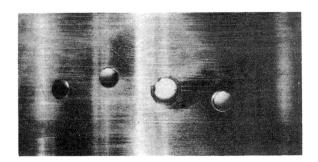

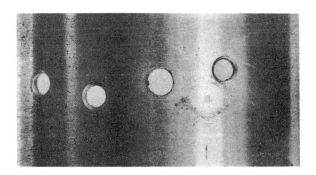

## Dirty assembly

### Characteristics

Localised overheating of the bearing surface and fatigue in extreme cases, sometimes in nominally lightly loaded areas.

### Causes

Entrapment of large particles of dirt (e.g. swarf), between bearing and housing, causing distortion of the shell, impairment of heat transfer and reduction of clearance (see next column).

## Dirty assembly

### Characteristics

Local areas of poor bedding on the back of the bearing shell, often around a 'hard' spot.

### Causes

Entrapment of dirt particles between bearing and housing. Bore of bearing is shown in previous column illustrating local overheating due to distortion of shell, causing reduction of clearance and impaired heat transfer.

## Cavitation erosion

### Characteristics

Removal of bearing material, especially soft overlays or whitemetal in regions near joint faces or grooves, leaving a roughened bright surface.

### Causes

Changes of pressure in oil film associated with interrupted flow.

## Discharge cavitation erosion

### Characteristics

Formation of pitting or grooving of the bearing material in a V-formation pointing in the direction of rotation.

### Causes

Rapid advance and retreat of journal in clearance during cycle. It is usually associated with the operation of a centrally grooved bearing at an excessive operating clearance.

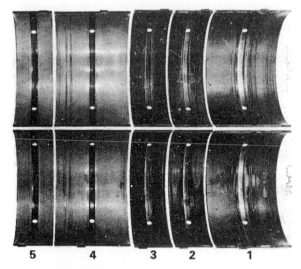

**5    4    3   2   1**

## Cavitation erosion

### Characteristics

Attack of bearing material in isolated areas, in random pattern, sometimes associated with grooves.

### Causes

Impact fatigue caused by collapse of vapour bubbles in oil film due to rapid pressure changes. Softer overlay (Nos 1, 2 and 3 bearings) attacked. Harder aluminium −20% tin (Nos 4 and 5 bearings) not attacked under these particular conditions.

## Corrosion

### Characteristics

Removal of lead phase from unplated copper–lead or lead–bronze, usually leading on to fatigue of the weakened material.

### Causes

Formation of organic acids by oxidation of lubricating oil in service. Consult oil suppliers; investigate possible coolant leakage into oil.

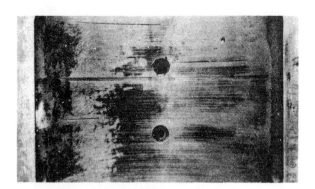

## Tin dioxide corrosion

### Characteristics

Formation of hard black deposit on surface of white-metal lining, especially in marine turbine bearings. Tin attacked, no tin-antimony and copper-tin constituents.

### Causes

Electrolyte (sea water) in oil.

## 'Sulphur' corrosion

### Characteristics

Deep pitting and attack or copper-base alloys, especially phosphor–bronze, in high temperature zones such as small-end bushes. Black coloration due to the formation of copper sulphide.

### Causes

Attack by sulphur-compounds from oil additives or fuel combustion products.

## 'Wire wool' damage

### Characteristics

Formation of hard black scab on whitemetal bearing surface, and severe machining away of journal in way of scab, as shown on the right.

### Causes

It is usually initiated by a large dirt particle embedded in the whitemetal, in contact with journal, especially chromium steel.

## 'Wire wool' damage

### Characteristics

Severe catastrophic machining of journal by 'black scab' formed in whitemetal lining of bearing. The machining 'debris' looks like wire wool.

### Causes

Self-propagation of scab, expecially with 'susceptible' journals steels, e.g. some chromium steels.

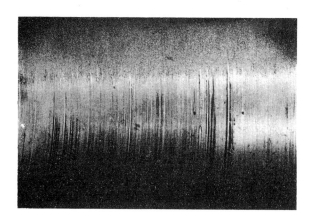

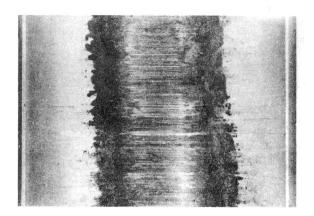

## Electrical discharge

### Characteristics

Pitting of bearing surface and of journal; may cause rapid failure in extreme cases

### Causes

Electrical currents from rotor to stator through oil film, often caused by faulty earthing.

## Fretting due to external vibration

### Characteristics

Pitting and pick-up on bearing surface.

### Causes

vibration transmitted from extgernal sources, causing damage while journal is stationary.

## Overheating

*Characteristics*

Extrusion and cracking, especially of whitemetal-lined bearings.

*Causes*

Operation at escessibe temperatures.

## Thermal cycling

*Characteristics*

Surface rumpling and grain-boundary cracking of tin-base whitemetal bearings.

*Causes*

Thermal cycling in service, causing plastic deformation, associated with the non-uniform thermal expansion of tin crystals.

## Faulty assembly

*Characteristics*

Localised fatigue or wiping in nominally lightly loaded areas.

*Causes*

Stagger at joint faces during assembly, due to excessive bolt clearances, or incorrect bolt disposition (bolts too far out).

## Faulty assembly

*Characteristics*

Overheating and pick-up at the sides of the bearings.

*Causes*

Incorrect grinding of journal radii, causing fouling at fillets.

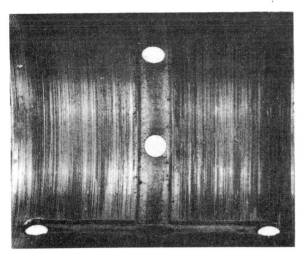

## Incorrect journal grinding

### Characteristics

Severe wiping and tearing-up of bearing surface.

## Causes

Too coarse a surface finish, or in the case of SG iron shafts, the final grinding of journal in wrong direction relative to rotation in bearing.

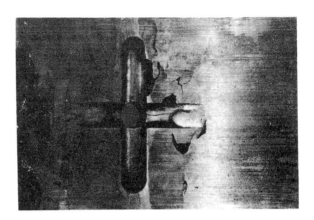

## Inadequate oil film thickness

### Characteristics

Fatigue cracking in proximity of a groove.

## Causes

Incorrect groove design, e.g. positioning a groove in the loaded area of the bearing.

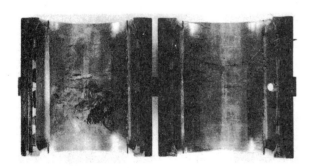

## Inadequate lubrication

### Characteristics

Seizure of bearing.

## Causes

Inadequate pump capacity or oil gallery or oilway dimensions. Blockage or cessation of oil supply.

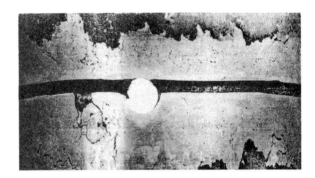

## Bad bonding

### Characteristics

Loss of lining, sometimes in large areas, even in lightly loaded regions, and showing full exposure of the backing material.

## Causes

Poor tinning of shells; incorrect metallurgical control of lining technique.

All photographs courtesy of Glacier Metal Co. Ltd

## FATIGUE FLAKE

### Characteristics

Flaking with conchoidal or ripple pattern extending evenly across the loaded part of the race.

### Causes

Fatigue due to repeated stressing of the metal. This is not a fault condition but it is the form by which a rolling element bearing should eventually fail. The multitude of small dents are caused by the debris and are a secondary effect.

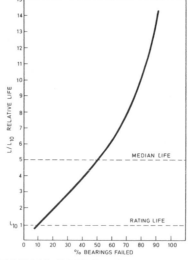

## EARLY FATIGUE FLAKE

### Characteristics

A normal fatigue flake but occurring in a comparatively short time. Appearance as for fatigue flake.

### Causes

Wide life-expectancy of rolling bearings. The graph shows approximate distribution for all types. Unless repeated, there is no fault. If repeated, load is probably higher than estimated; check thermal expansion and centrifugal loads.

## ATMOSPHERIC CORROSION

### Characteristics

Numerous irregular pits, reddish brown to dark brown in colour. Pits have rough irregular bottoms.

### Causes

Exposure to moist conditions, use of a grease giving inadequate protection against water corrosion.

## ROLLER STAINING

### Characteristics

Dark patches on rolling surfaces and end faces of rollers in bearings with yellow metal cages. The patches usually conform in shape to the cage bars.

### Causes

Bi-metallic corrosion in storage. May be due to poor storage conditions or insufficient cleaning during manufacture. Special packings are available for severe conditions. Staining, as shown, can be removed by the manufacturer, to whom the bearing should be returned.

## BRUISING (OR TRUE BRINELLING)

### Characteristics

Dents or grooves in the bearing track conforming to the shape of the rolling elements. Grinding marks not obliterated and the metal at the edges of the dents has been slightly raised.

### Causes

The rolling elements have been brought into violent contact with the race; in this case during assembly using impact.

## FALSE BRINELLING

### Characteristics

Depressions in the tracks which may vary from shallow marks to deep cavities. Close inspection reveals that the depressions have a roughened surface texture and that the grinding marks have been removed. There is usually no tendency for the metal at the groove edges to have been displaced.

### Causes

Vibration while the bearing is stationary or a small oscillating movement while under load.

## FRACTURED FLANGE
### Characteristics
Pieces broken from the inner race guiding flange. General damage to cage and shields.

### Causes
Bad fitting. The bearing was pressed into housing by applying load to the inner race causing cracking of the flange. During running the cracks extended and the flange collapsed. A bearing must never be fitted so that the fitting load is transmitted via the rolling elements.

## OUTER RACE FRETTING
### Characteristics
A patchy discoloration of the outer surface and the presence of reddish brown debris ('cocoa'). The race is not softened but cracks may extend inwards from the fretted zone.

### Causes
Insufficient interference between race and housing. Particularly noticeable with heavily loaded bearings having thin outer races.

## INNER RACE FRETTING
### Characteristics
Heavy fretting of the shaft often with gross scalloping; presence of brown debris ('cocoa'). Inner race may show some fretting marks.

### Causes
Too little interference, often slight clearance, between the inner race and the shaft combined with heavy axial clamping. Axial clamping alone will not prevent a heavily loaded inner race precessing slowly on the shaft.

## INNER RACE SPINNING
### Characteristics
Softening and scoring of the inner race and the shaft, overheating leading to carbonisation of lubricant in severe cases, may lead to complete seizure.

### Causes
Inner race fitted with too little interference on shaft and with light axial clamping.

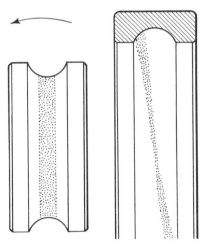

## SKEW RUNNING MARKS
### Characteristics
The running marks on the stationary race are not parallel to the faces of the race. In the figure the outer race is stationary.

### Causes
Misalignment. The bearing has not failed but may do so if allowed to continue to run out of line.

## UNEVEN FATIGUE
### Characteristics
Normal fatigue flaking but limited to, or much more severe on, one side of the running track.

### Causes
Misalignment.

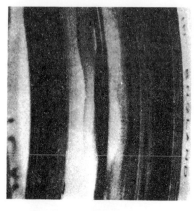

## UNEVEN WEAR MARKS
### Characteristics
The running or wear marks have an uneven width and may have a wavy outline instead of being a uniform dark band.

### Causes
Ball skidding due to a variable rotating load or local distortion of the races.

## ROLLER END COLLAPSE
### Characteristics
Flaking near the roller-end radius at one end only. Microscopic examination reveals roundish smooth-bottomed pits.

### Causes
Electrical damage with some misalignment. If the pits are absent then the probable cause is roller end bruising which can usually be detected on the undamaged shoulder. Although misalignment accentuates this type of damage it has rarely been proved to be the sole cause.

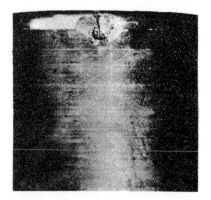

## ROLLER END CHIPPING
### Characteristics
A collapse of the material near the corner radii of the roller. In this instance chipping occurred simultaneously at opposite ends of the roller. A well-defined sub-surface crack can be seen.

### Causes
Subcutaneous inclusions running the length of the roller. This type of failure is more usually found in the larger sizes of bearing.

Chipping at one end only may be caused by bruising during manufacture, or by electrical currents, and accentuated by misalignment.

## ROLLER PEELING
### Characteristics
Patches of the surface of the rollers are removed to a depth of about 0.0005 in.

### Causes
This condition usually follows from an initial mild surface damage such as light electrical pitting; this could be confirmed by microscopic examination. It has also been observed on rollers which were slightly corroded before use.

If the cause is removed this damage does not usually develop into total failure.

## ROLLER BREAKAGE
### Characteristics
One roller breaks into large fragments which may hold together. Cage pocket damaged.

### Causes
Random fatigue. May be due to faults or inclusions in the roller material. Replacement bearing usually performs satisfactorily.

## MAGNETIC DAMAGE
### Characteristics
Softening of the rotating track and rolling elements leading to premature fatigue flaking.

### Causes
Bearing has been rotating in a magnetic field (in this case, 230 kilolines ($230 \times 10^{-5}$ Wb), 300 rev/min, 860 h).

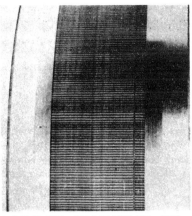

## LADDER MARKING OR WASHBOARD EROSION

*Characteristics*

A regular pattern of dark and light bands which may have developed into definite grooves. Microscopic examination shows numerous small, almost round, pits.

*Causes*

An electric current has passed across the bearing; a.c. or d.c. currents will cause this effect which may be found on either race or on the rolling elements.

## GREASE FAILURE

*Characteristics*

Cage pockets and rims worn. Remaining grease dry and hard; bearing shows signs of overheating.

*Causes*

Use of unsuitable grease. Common type of failure where temperatures are too high for the grease in use.

## MOLTEN CAGE

*Characteristics*

Cage melted down to the rivets, inner race shows temper colours.

*Causes*

Lubrication failure on a high-speed bearing. In this case an oil failure at 26 000 rev/min. In a slower bearing the damage would not have been so localised.

## OVERHEATING

*Characteristics*

All parts of the bearing are blackened or show temper colours. Lubricant either absent or charred. Loss of hardness on all parts.

*Causes*

Gross overheating. Mild overheating may only show up as a loss of hardness.

## SMEARING

*Characteristics*

Scuff marks, discoloration and metal transfer on non-rolling surfaces. Usually some loss of hardness and evidence of detrioration of lubricant. Often found on the ends of rollers and the corresponding guide face on the flanges.

*Causes*

Heavy loads and/or poor lubrication.

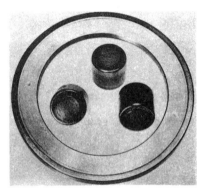

## ABRASIVE WEAR

*Characteristics*

Dulling of the working surfaces and the removal of metal without loss of hardness.

*Causes*

Abrasive particles in the lubricant, usually non-metallic.

# Gear failures

Gear failures rarely occur. A gear pair has not failed until it can no longer be run. This condition is reached when (*a*) one or more teeth have broken away, preventing transmission of motion between the pair or (*b*) teeth are so badly damaged that vibration and noise are unacceptable when the gears are run.

By no means all tooth damage leads to failure and immediately it is observed, damaged teeth should be examined to determine whether the gears can safely continue in service.

## SURFACE FATIGUE

This includes case exfoliation in skin-hardened gears and pitting which is the commonest form of damage, especially with unhardened gears. Pitting, of which four types are distinguished, is indicated by the development of relatively smooth-bottomed cavities generally on or below the pitch line. In isolation they are generally conchoidal in appearance but an accumulation may disguise this.

### Case exfoliation

*Case exfoliation on a spiral bevel pinion*

#### *Characteristics*

Appreciable areas of the skin on surface hardened teeth flake away from the parent metal in heavily loaded gears. Carburised and hardened, nitrided and induction hardened materials are affected.

#### *Causes*

Case exfoliation often indicates a hardened skin that is too thin to support the tooth load. Cracks sometimes originate on the plane of maximum Hertzian shear stress and subsequently break out to the surface, but more often a surface crack initiates the damage. Another possible reason for case exfoliation is the high residual stress resulting from too severe a hardness gradient between case and core. Exfoliation may be prevented by providing adequate case depth and tempering the gear material after hardening.

### Initial or arrested pitting

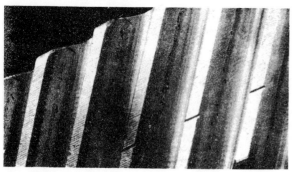

*Initial or arrested pitting on a single helical gear*

#### *Characteristics*

Initial pitting usually occurs on gears that are not skin hardened. It may be randomly distributed over the whole tooth flank, but more often is found around the pitch line or in the dedendum. Single pits rarely exceed 2 mm across and pitting appears in the early running life of a gear.

#### *Causes*

Discrete irregularities in profile or surface asperities are subjected to repeated overstress as the line of contact sweeps across a tooth to produce small surface cracks and clefts. In the dedendum area the oil under the high pressure of the contact can enter these defects and extend them little by little, eventually reaching the surface again so that a pit is formed and a small piece of metal is dislodged. Removal of areas of overstress in this way spreads the load on the teeth to a level where further crack or cleft formation no longer occurs and pitting ceases.

## Progressive or potentially destructive pitting

PITTING ON SOME TEETH
IS CONTINUOUS AND
QUITE DEEP

*Progressive pitting on single helical gear teeth*

### Characteristics

Pits continue to form with continued running, especially in the dedendum area. Observation on marked teeth will indicate the rate of progress which may be intermittent. A rapid increase, particularly in the root area, may cause complete failure by increasing the stress there to the point where large pieces of teeth break away.

### Causes

Essentially the gear material is generally overstressed, often by repeated shock loads. With destructive pitting the propagating cracks branch at about the plane of maximum Hertzian shear stress; one follows the normal initial pitting process but the other penetrates deeper into the metal.

Remedial action is to remove the cause of the overload by correcting alignment or using resilient couplings to remove the effect of shock loads. The life of a gear based on surface fatigue is greatly influenced by surface stress. Thus, if the load is carried on only half the face width the life will only be a small fraction of the normal value. In slow and medium speed gears it may be possible to ameliorate conditions by using a more viscous oil, but this is generally ineffective with high speed gears.

In skin-hardened gears pits of very large area resembling case exfoliation may be formed by excessive surface friction due to the use of an oil lacking sufficient viscosity.

## Dedendum attrition

*Dedendum attrition on a large single helical gear*

### Characteristics

The dedendum is covered by a large number of small pits and has a matt appearance. Both gears are equally affected and with continued running the dedenda are worn away and a step is formed at the pitch line to a depth of perhaps 0.5 mm. The metal may be detached as pit particles or as thin flakes. The wear may cease at this stage but may run in cycles, the dedenda becoming smooth before pitting restarts. If attrition is permitted to continue vibration and noise may become intolerable. Pitting may not necessarily be present in the addendum.

### Causes

The cause of this type of deterioration is not fully understood but appears to be associated with vibration in the gear unit. Damage may be mitigated by the use of a more viscous oil.

## Micro-pitting

### Characteristics

Found predominantly on the dedendum but also to a considerable extent on the addendum of skin-hardened gears. To the naked eye affected areas have a dull grey, matt or 'frosted' appearance but under the microscope they are seen to be covered by a myriad of tiny pits ranging in size from about 0.03 to 0.08 mm and about 0.01 mm deep.

Depending on the position of the affected areas, micro-pitting may be corrective, especially with helical gears.

### Causes

Overloading of very thin, brittle and super-hard surface layers, as in nitrided surfaces, or where a white-etching layer has formed, by normal and tangential loads. Coarse surface finishes and low oil viscosity can be predisposing factors. In some cases it may be accelerated by unsuitable load-carrying additives in the oil.

## SMOOTH CHEMICAL WEAR

Can arise where gears using extreme pressure oil run under sustained heavy loads, at high temperatures.

## Smooth chemical wear

Hypoid pinion showing smooth chemical wear

### Characteristics

The working surfaces of the teeth, especially of the pinion, are worn and have a burnished appearance.

### Causes

Very high surface temperatures cause the scuff resistant surface produced by chemical reaction with the steel to be removed and replaced very rapidly. The remedies are to reduce the operating temperatures, to reduce tooth friction by using a more viscous oil and to use a less active load-carrying additive.

## SCUFFING

Scuffing occurs at peripheral speeds above about 3 m/s and is the result of either the complete absence of a lubricant film or its disruption by overheating. Damage may range from a lightly etched appearance (slight scuffing) to severe welding and tearing of engaging teeth (heavy scuffing). Scuffing can lead to complete destruction if not arrested.

### Light scuffing

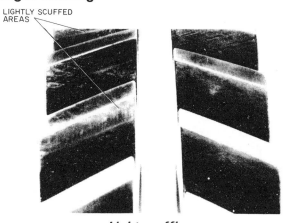

LIGHTLY SCUFFED AREAS

*Light scuffing*

### *Characteristics*

Tooth surfaces affected appear dull and slightly rough in comparison with unaffected areas. Low magnification of a scuffed zone reveals small welded areas subsequently torn apart in the direction of sliding, usually at the tip and root of the engaging teeth where sliding speed is a maximum.

### *Causes*

Disruption of the lubricant film occurs when the gear tooth surfaces reach a critical temperature associated with a particular oil and direct contact between the sliding surfaces permits discrete welding to take place. Low viscosity plain oils are more liable to permit scuffing than oils of higher viscosity. Extreme pressure oils almost always prevent it.

### Heavy scuffing

*Heavy scuffing on a case hardened hypoid wheel*

### *Characteristics*

Tooth surfaces are severely roughened and torn as the result of unchecked adhesive wear.

### *Causes*

This is the result of maintaining the conditions that produced light scuffing. The temperature of the contacting surfaces rises so far above the critical temperature for the lubricant that continual welding and tearing of the gear material persists.

Spur, helical and bevel gears, may show so much displacement of the metal that a groove is formed along the pitch line of the driving gear and a corresponding ridge on that of the driven gear. It may be due to the complete absence of lubricant, even if only temporarily. Otherwise, the use of a more viscous oil, or one with extreme pressure properties is called for.

## GENERAL COMMENTS ON GEAR TOOTH DAMAGE

Contact marking is the acceptance criterion for all toothed gearing, and periodic examination of this feature until the running pattern has been established, is the most satisfactory method of determining service performance. It is therefore advisable to look at the tooth surfaces on a gear pair soon after it has been run under normal working conditions. If any surface damage is found it is essential that the probable cause is recognised quickly and remedial action taken if necessary, before serious damage has resulted. Finding the principal cause may be more difficult when more than one form of damage is present, but it is usually possible to consider each characteristic separately.

The most prolific sources of trouble are faulty lubrication and misalignment. Both can be corrected if present, but unless scuffing has occurred, further periodic observation of any damaged tooth surfaces should be made before taking action which may not be immediately necessary.

## ABRASIVE WEAR

During normal operation, engaging gear teeth are separated from one another by a lubricant film, commonly about 0.5μm thick. Where both gears are unhardened and abrasive particles dimensionally larger than the film thickness contaminate the lubricant, especially if it is a grease, both sets of tooth surfaces are affected (three-body abrasion). Where one gear has very hard tooth surfaces and surface roughness greater than the film thickness, two-body abrasive wear occurs and the softer gear only becomes worn. For example, a rough case-hardened steel worm mating with a bronze worm wheel, or a rough steel pinion engaging a plastic wheel.

### Foreign matter in the lubricant

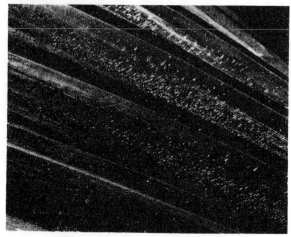

*Effect of foreign matter in lubricant*

### *Characteristics*

Grooves are cut in the tooth flanks in the direction of sliding and their size corresponds to the size of the contaminant present. Displaced material piles up along the sides of a groove or is removed as a fine cutting. Usually scratches are short and do not extend to the tooth tips.

### *Causes*

The usual causes of three-body abrasion are gritty materials falling into an open gear unit or, in an enclosed unit, inadequate cleaning of the gear case and oil supply pipes of such materials as casting sand, loose scale, shot-blast grit, etc.

### Attrition caused by fine foreign matter in oil

*Spur gear virtually destroyed by foreign matter in the oil*

### *Characteristics*

These are essentially similar to lapping. Very fine foreign matter suspended in a lubricant can pass through the gear mesh with little effect when normal film lubrication prevails. Unfavourable conditions permit abrasive wear; tooth surfaces appear dull and scratched in the direction of sliding. If unchecked, destruction of tooth profiles results from the lapping.

### *Causes*

The size of the foreign matter permits bridging through the oil film. Most frequently, the origin of the abrasive material is environmental. Both gears and bearings suffer and systems should be cleaned, flushed, refilled with clean oil and protected from further contamination as soon as possible after discovery.

## TOOTH BREAKAGE

If a whole tooth breaks away the gear has failed but in some instances a corner of a tooth may be broken and the gear can continue to run. The cause of a fracture should influence an assessment of the future performance of a gear.

### Brittle fracture resulting from high shock load

*Brittle fracture on spiral bevel wheel teeth*

*Characteristics*

More than one tooth may be affected. With hard steels the entire fracture surface appears to be granular denoting a brittle fracture. With more ductile materials the surface has a fibrous and torn appearance.

*Causes*

A sudden and severe shock load has been applied to one or other member of a gear pair which has greatly exceeded the impact characteristics of the material. A brittle fracture may also indicate too low an Izod value in the gear material, though this is a very rare occurrence. A brittle fracture in bronze gears indicates the additional effect of overheating.

### Tooth end and tip loading

LIGHT SCUFFING

*Tooth end and tip loading*

*Characteristics*

Spiral bevel and hypoid gears are particularly liable to heel end tooth breakage and other types of skin hardened gears may have the tooth tips breaking away. Fractured surfaces often exhibit rapid fatigue characteristics.

*Causes*

The immediate cause is excessive local loading. This may be produced by very high transmitted torque, incorrect meshing or insufficient tip relief.

23

## Impact or excessive loading causing fatigue fracture

*Slow fatigue on a through-hardened helical wheel*

### Characteristics

Often exhibits cracks in the roots on the loaded side of a number of teeth. If teeth have broken out the fracture surfaces show two phases; a very fine-grained, silky, conchoidal zone starting from the loaded side followed, where the final failure has suddenly occurred, by a coarse-grained brittle fracture.

### Causes

The loading has been so intense as to exceed the tensile bending stress limit resulting in root cracking. Often stress-raisers in the roots such as blowholes, bruises, deep machining marks or non-metallic inclusions, etc. are involved. If the excessive loading continues the teeth will break away by slow fatigue and final sudden fracture.

## Fatigue failure resulting from progressive pitting

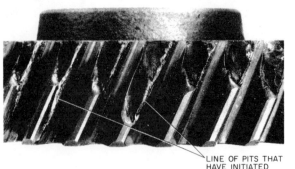

LINE OF PITS THAT HAVE INITIATED FRACTURES

*Fatigue failure from progressive pitting*

### Characteristics

Broken tooth surfaces exhibit slow fatigue markings, with the origin of the break at pits in the dedendum of the affected gear.

### Causes

Progressive pitting indicates that the gears are being run with a surface stress intensity above the fatigue limit. Cracks originating at the surface continue to penetrate into the material.

## PLASTIC DEFORMATION

Plastic deformation occurs on gear teeth due to the surface layers yielding under heavy loads through an intact oil film. It is unlikely to occur with hardnesses above HV 350.

### Severe plastic flow in steel gears

*Severe plastic flow in helical gears*

### Characteristics

A flash or knife-edge is formed on the tips of the driving teeth often with a hollow at the pitch cylinder and a corresponding swelling on the driven teeth. The ends of the teeth can also develop a flash and the flanks are normally highly burnished.

### Causes

The main causes are heavy steady or repeated shock loading which raises the surface stress above the elastic limit of the material, the surface layers being displaced while in the plastic state, especially in the direction of sliding. Since a work-hardened skin tends to develop, the phenomenon is not necessarily detrimental, especially in helical gears, unless the tooth profiles are severely damaged. A more viscous oil is often advantageous, particularly with shock-loading, but the best remedy is to reduce the transmitted load, possibly by correcting the alignment.

## CASE CRACKING

With correctly manufactured case hardened gears case cracking is a rare occurrence. It may appear as the result of severe shock or excessive overload leading to tooth breakage or as a condition peculiar to worm gears.

### Heat/load cracking on worms

RADIAL CRACKS IN POLISHED
CONTACT ZONE

*Heat/load cracking on a worm wheel*

### Characteristics

On extremely heavily loaded worms the highly polished contact zone may carry a series of radial cracks. Spacing of the cracks is widest where the contact band is wide and they are correspondingly closer spaced as the band narrows. Edges rarely rise above the general level of the surface.

### Causes

The cracks are thought to be the result of high local temperatures induced by the load. Case hardened worms made from high core strength material (En39 steel) resist this type of cracking.

## FAILURES OF PLASTIC GEARS

Gears made from plastic materials are meshed with either another plastic gear or more often, with a cast iron or steel gear; non ferrous metals are seldom used. When applicable, failures generally resemble those described for metal gears.

Severe plastic flow, scoring and tooth fracture indicate excessive loading, possibly associated with inadequate lubrication. Tempering colours on steel members are the sign of unsatisfactory heat dispersal by the lubricant.

Wear on the metallic member of a plastic/metal gear pair usually suggests the presence of abrasive material embedded in the plastic gear teeth. This condition may derive from a dusty atmosphere or from foreign matter carried in the lubricant.

When the plastic member exhibits wear the cause is commonly attributable to a defective engaging surface on the metallic gear teeth. Surface texture should preferably not be rougher than $16\mu$in ($0.4\mu$m) cla.

## PISTON PROBLEMS

Piston problems usually arise from three main causes and these are:
1. Unsatisfactory rubbing conditions between the piston and the cylinder.
2. Excessive operating temperature, usually caused by inadequate cooling or possibly by poor combustion conditions.
3. Inadequate strength or stiffness of the piston or associated components at the loads which are being applied in operation.

### Skirt scratching and scoring

*Characteristics*

The piston skirt shows axial scoring marks predominantly on the thrust side. In severe cases there may be local areas showing incipient seizure.

*Causes*

Abrasive particles entering the space between the piston and cylinder. This can be due to operation in a dusty environment with poor air filtration. Similar damage can arise if piston ring scuffing has occurred since this can generate hard particulate debris. More rarely the problem can arise from an excessively rough cylinder surface finish.

### Piston skirt seizure

*Characteristics*

Severe scuffing damage, particularly on the piston skirt but often extending to the crown and ring lands. The damage is often worse on the thrust side.

*Causes*

Operation with an inadequate clearance between the piston and cylinder. This can be associated with inadequate cooling or a poor piston profile. Similar damage could also arise if there was an inadequate rate of lubricant feed up the bore from crankshaft bearing splash.

### Piston crown and ring land damage

*Characteristics*

The crown may show cracking and the crown land and lands between the rings may show major distortion, often with the ring ends digging in to the lands.

*Causes*

Major overheating caused by poor cooling and in diesel engines defective injectors and combustion. The problem may arise from inadequate cylinder coolant flow or from the failure of piston cooling arising from blocked oil cooling jets.

*Skirt scratching*

*Skirt seizure*

## Misaligned pistons

### *Characteristics*

The bedding on the skirt is not purely axial but shows diagonal bedding.

### *Causes*

Crankshaft deflections or connecting rod bending. Misalignment of rod or gudgeon pin bores.

## Cracking inside the piston

### *Characteristics*

Cracks near the gudgeon pin bosses and behind the ring grooves.

### *Causes*

Inadequate gudgeon pin stiffness can cause cracking in adjacent parts of the piston, or parts of the piston cross section may be of inadequate area.

*Diagonal skirt bedding*

## RING PROBLEMS

The most common problem with piston rings is scuffing of their running surfaces. Slight local scuffing is not uncommon in the first 20 to 50 hours of running from new when the rings are bedding in to an appropriate operating profile. However the condition of the ring surfaces should progressively improve and scuffing damage should not spread all round the rings.

## Scuffing of cast iron rings

### *Characteristics*

Local zones around the ring surface where there are axial dragging marks and associated surface roughening. Detailed examination often shows thin surface layers of material with a hardness exceeding 1000 Hv and composed of non-etching fine grained martensite (white layer).

### *Causes*

Can arise from an unsuitable initial finish on the cylinder surface. It can also arise if the rings tend to bed at the top of their running surface due to unsuitable profiling or from thermal distortion of the piston.

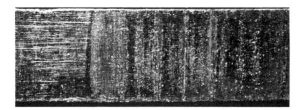

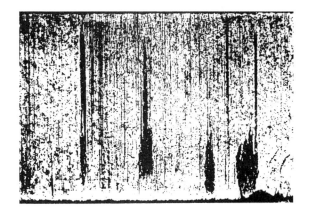

*Scuffed cast iron rings*

## Scuffing of chromium plated piston rings

### Characteristics

The presence of dark bands running across the width of the ring surface usually associated with transverse circumferential cracks. In severe cases portions of the chromium plating may be dragged from the surface.

### Causes

Unsuitable cylinder surface finish or poor profiling of the piston rings. Chromium plated top rings need to have a barelled profile as installed to avoid hard bedding at the edges.

In some cases the problem can also arise from poor quality plating in which the plated surface is excessively rough or globular and can give local sharp areas on the ring edges after machining.

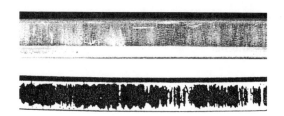

*Scuffed chromium plated rings*

*Severely damaged chromium plate*

MACHINED SURFACE
OF THE RING

CHAMFER AT EDGE
OF THE PLATING.
GLOBULAR FINISH CAN
CREATE LOCAL SHARP EDGES.

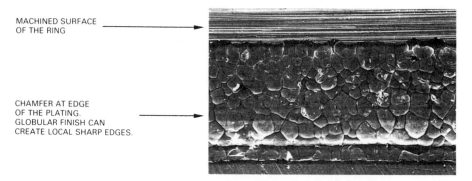

*The edge of a piston ring*

## Rings sticking in their grooves

### Characteristics

The rings are found to be fixed in their grooves or very sluggish in motion. There may be excessive blow by or oil consumption.

### Causes

The ring groove temperatures are too high due to conditions of operation or poor cooling. The use of a lubricating oil of inadequate quality can also aggravate the problem.

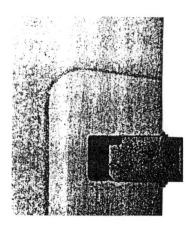

*A stuck piston ring*

## CYLINDER PROBLEMS

Problems with cylinders tend to be of three types:
1. Running in problems such as bore polishing or in some cases scuffing.
2. Rates of wear in service which are high and give reduced life.
3. Other problems such as bore distortion arising from the engine design or cavitation erosion damage of the water side of a cylinder liner, which can penetrate through to the bore.

## Bore polishing

### Characteristics

Local areas of the bore surface become polished and oil consumption and blow by tend to increase because the piston rings do not then bed evenly around the bore. The polished areas can be very hard thin, wear-resistant 'white' layers.

### Causes

The build up of hard carbon deposits on the top land of the piston can rub away local areas of the bore surface and remove the controlled surface roughness required to bed in the piston rings.

If there is noticeable bore distortion from structural deflections or thermal effects, the resulting high spots will be preferentially smoothed by the piston rings.

The chemical nature of the lubricating oil can be a significant factor in both the hard carbon build-up and in the polishing action.

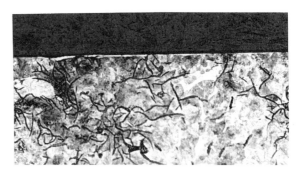

*Bore polishing*

## High wear of cast iron cylinders

### Characteristics

Cylinder liners wear in normal service due to the action of fine abrasive particles drawn in by the intake air. The greatest wear occurs near to the TDC position of the top ring.

Corrosion of a cast iron bore surface can however release hard flake-like particles of iron carbide from the pearlite in the iron. These give a greatly increased rate of abrasive wear.

### Causes

Inadequate air filtration when engines are operated in dusty environments.

Engines operating at too low a coolant temperature, i.e. below about 80°C, since this allows the internal condensation of water vapour from the combustion process, and the formation of corrosion pits in the cylinder surface.

*Corrosion of a cast iron bore*

## High wear of chromium plated cylinders

### Characteristics

An increasing rate of wear with operating time associated with the loss of the surface profiling which provides a dispersed lubricant supply. The surface becomes smooth initially and then scuffs because of the unsatisfactory surface profile. This then results in a major increase in wear rate.

### Causes

High rates of abrasive particle ingestion from the environment can cause this problem. A more likely cause may be inadequate quality of chromium plating and its finishing process aimed at providing surface porosity. Some finishing processes can leave relatively loose particles of chromium in the surface which become loose in service and accelerate the wear process.

## Bore scuffing

### Characteristics

Occurs in conjunction with piston ring scuffing. The surface of the cylinder shows areas where the metal has been dragged in an axial direction with associated surface roughening.

### Causes

The same as for piston ring scuffing but in addition the problem can be accentuated if the metalurgical structure of the cylinder surface is unsatisfactory.

In the case of cast iron the material must be pearlitic and should contain dispersed hard constituents derived from phosphorous, chromium or vanadium constituents. The surface finish must also be of the correct roughness to give satisfactory bedding in of the piston rings.

In the case of chromium plated cylinder liners it is essential that the surface has an undulating or grooved profile to provide dispersed lubricant feeding to the surface.

## Cavitation erosion of cylinder liners

### Characteristics

If separate cylinder liners are used with coolant in contact with their outside surface, areas of cavitation attack can occur on the outside. The material removal by cavitation continues and eventually the liner is perforated and allows the coolant to enter the inside of the engine.

### Causes

Vibration of the cylinder liner under the influence of piston impact forces is the main cause of this problem but it is accentuated by crevice corrosion effects if the outside of the liner has dead areas away from the coolant flow.

*Abrasive turn round marks at TDC*

*A chromium plated liner which has scuffed after losing its surface profiling by wear*

## ROTARY MECHANICAL SEALS

### Table 6.1 Common failure mechanisms of mechanical seals

| Special conditions | Likely symptoms | | | Seizure | Failure mechanism | Remedy |
|---|---|---|---|---|---|---|
| | High leakage | High friction | High wear | | | |
| **OPERATING CONDITIONS** | | | | | | |
| Speed of sliding high | | X | X | X | Excessive frictional heating, film vaporises | Provide cooling |
| " | X | | | | Thermal stress cracking of the face (Figure 6.1) | Use material with higher conductivity or higher tensile strength |
| " | X | X | X | X | Thermal distortion of seal (Figure 6.2) | Provide cooling |
| Speed of sliding low | | X | X | X | Poor hydrodynamic lubrication, solid contact | Use face with good boundary lubrication capacity |
| Appreciable vibration present | X | | X | X | Face separation unstable | Try to reduce vibration, avoid bellows seals, fit damper |
| Low pressure differential | (X) | | | | Fluid pumped by seal against pressure | Try reversing seal to redirect flow |
| High pressure differential | | X | X | X | Hydrodynamic film overloaded | Modify area ratio of seal to reduce load |
| " | X | X | X | X | Seal or housing distorting (Figure 6.2) | Stiffen seal and/or housing |
| Sterilisation or cleaning cycle used | X | | | | High temperature or solvents incompatible with seal materials, especially rubbers | Use compatible materials |
| Exposure to sunlight, ozone, radiation | X | | | | Seal materials (rubber) fail | Protect seal from exposure, consider other materials |
| **FLUID** | | | | | | |
| Viscosity of fluid high | | X | X | X | Excessive frictional heating, film vaporises | Provide cooling |
| | X | X | X | X | Excessive frictional heating, seal distorts | Provide cooling |
| Viscosity of fluid low | | X | X | X | Poor hydrodynamic lubrication, solid contact | Use faces with good boundary lubrication capacity |
| Lubricity of fluid poor | X | X | X | X | Surfaces seize or 'pick-up' | Use faces with good boundary lubrication capacity |
| Abrasives in fluid | X | | X | | Solids in interface film | Circulate clean fluid round seal |
| Crystallisable fluid | X | | X | | Crystals form at seal face | Raise temperature or flush fluid outside seal |
| Polymerisable fluid | X | | X | | Solids form at seal face | Raise temperature or flush fluid outside seal |
| Ionic fluid, e.g. salt solutions | X | X | X | X | Corrosion damages seal faces | Select resistant materials |
| Non-Newtonian fluid, e.g. suspension, colloids, etc. | (X) | | | | Fluid behaves unpredictably, leakage may be reversed | Try reversing seal to redirect flow in acceptable direction |
| **DESIGN** | | | | | | |
| Auxiliary cooling, flushing, etc. | X | X | X | X | Stoppage in auxiliary circuit | Overhaul auxiliaries |
| Double seals | X | X | X | X | Pressure build-up between seals if there is no provision for pressure control | Provide pressure control |
| Housing flexes due to pressure or temperature changes | X | X | X | X | Seal faces out of alignment, non-uniform wear | Stiffen housing and/or mount seal flexibly |
| Seal face flatness poor | X | | | | Excessive seal gap (Figure 6.3) | Lap faces flatter |
| Seal faces rough | X | X | X | X | Asperities make solid contact | Lap or grind faces |
| Bellows type seal | X | | | | Floating seal member vibrates | Fit damping device to bellows |

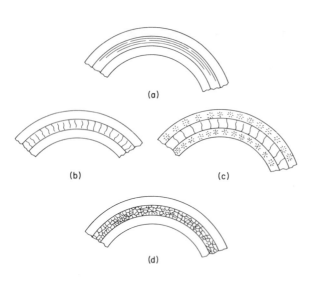

**Figure 6.1 Mechanical seal faces after use:** *(a) normal appearance, some circumferential scoring; (b) parallel radial cracks; (c) radial cracks with blisters; (d) surface crazing; (b) – (d) are due to overheating, particularly characteristic of ceramic seal faces*

**Figure 6.2 Tungsten carbide mechanical seal face showing symmetrical surface polishing characteristic of mild hydraulic or thermal distortion**

**Figure 6.3 Tungsten carbide mechanical seal face showing localised polishing due to lack of flatness; this seal leaked badly.** *The inset illustrates a typical non-flat seal face viewed in sodium light using an optical flat to give contour lines at 11 micro-inch increments of height*

## RUBBER SEALS OF ALL TYPES

### Table 6.2 Common failure mechanisms of rubber seals

| Symptoms | Cause | Remedy |
|---|---|---|
| Rubber brittle, possibly cracked, seal leaks | Rubber ageing. Exposure to ozone/sunlight. Overheated due to high fluid temperature or high speed (Figure 6.4) | Renew seal, consider change of rubber compound; consider improving seal environmental or operating conditions |
| Rubber softened, possibly swollen | Rubber incompatible with sealed fluid (Figure 6.5) | Change rubber compound or fluid |
| Seal motion irregular, jerky, vibration, or audible squeal | 'Stick-slip' (Figure 6.6) | Higher or lower speed may avoid problem; change fluid temperature; change rubber compound |
| Seal friction very high on starting | 'Stiction', i.e. static friction is time dependent and much higher than kinetic friction (Figure 6.7) | Probably inevitable in some degree, time effect slowed by softer rubber and/or more viscous fluid |
| Seal permanently deformed | 'Permanent set', a characteristic of rubbers, some more than others (Figure 6.4) | Change rubber compound |

**Figure 6.4 Rubber O-ring failure due to overheating.** *The brittle fracture is due to hardening of the originally soft rubber and the flattened appearance is a typical example of compression set*

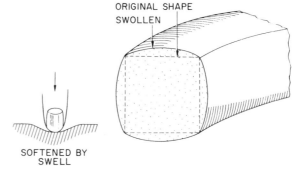

**Figure 6.5 Rubber-fluid incompatibility**

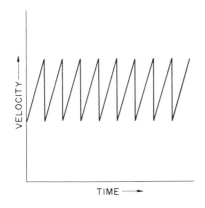

**Figure 6.6 'Stick-slip'**

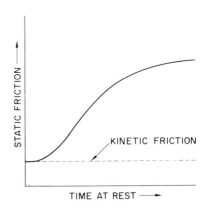

**Figure 6.7 'Stiction'**

# Seal failures

## O-Rings, Rectangular Rubber Rings, etc.

### Table 6.3 Common failure mechanisms

| Symptoms | Cause | Remedy |
|---|---|---|
| Fine circumferential cut set back slightly from sliding contact zone. Rubber 'nibbled' on one side. Ring completely ejected from its groove | Extrusion damage (Figure 6.8) | Reduce back clearance, check concentricity of parts; fit back-up ring; use reinforced seal; use harder rubber |
| Wear, not restricted to sliding contact zone. Partial or total fracture | Ring rolling or twisting in groove | Replace O-ring by rectangular section ring or a lobed type ring |

**Figure 6.8 A rectangular-section rubber seal ring showing extrusion damage.** *Where damage is less severe (r.h.s.) only a knife-cut is visible, but material has been nibbled away where extrusion was severe. The circumferential variation of the damage indicates eccentricity of the sealed components*

**Figure 6.9 Wear failure of a rubberised-fabric square-back U-ring due to inadequate lubrication when sealing distilled water. Friction was also bad**

## Reciprocating Seals

### Table 6.4 Common failure mechanisms

| Symptoms | Cause | Remedy |
|---|---|---|
| Excessive wear and/or high friction | Poor lubrication. (Figure 6.9) Seal overloaded | If multiple seals are in use replace with single seal. Heavy duty seal may cure overloading. With aqueous fluids leather may be better than rubber |
| Non-uniform wear circumferentially | Dirt ingress Deposits on rod Side load | Fit wiper or scraper " Check bearings |

## Rotary Lip Seals

### Table 6.5 Common failure mechanisms

| Symptoms | Cause | Remedy |
|---|---|---|
| Excessive leakage | Damaged lip. Machining has left a spiral lead on shaft | Check for damage and cause, e.g. contact with splines or other rough surface during assembly or careless handling/storage. Use mandrel for assembly. Eliminate shaft lead |
| | Unsuitable shaft surface | Finish the surface to $R_A$ 0.1–0.5$\mu$m |
| Lip cracked in places | Excessive speed. Poor lubrication. Hot environment | Consider alternative rubber compounds. Improve lubrication. Reduce environmental temperature |

## PACKED GLANDS

### Table 6.6 Common failure mechanisms of packed glands

| Symptoms | Cause | Remedy |
|---|---|---|
| Packing extruded into clearance between shaft and housing or gland follower (Figure 6.10) | Designed clearance excessive or parts worn by abrasives or shaft bearings inadequate | Reduce clearances, check bearings |
| Leakage along outside of gland follower (Figure 6.11) | Packing improperly fitted or housing bore condition bad | Repack with care after checking bore condition |
| Used packing scored on outside surface, possibly leakage along outside of gland follower | Packing rotating with shaft due to being undersized | Check dimensions of housing and packing |
| Packing rings near gland follower very compressed (Figure 6.12) or rings embedded into each other | Packing incorrectly sized | Repack with accurately sized packing rings |
| Bore of used packing charred or blackened possibly shaft material adhering to packing | Lubrication failure | Change packing to one with more suitable lubricants or fit lantern ring with lubricant feed |
| Shaft badly worn along its length (Figure 6.13) | Lubrication failure or abrasive solids present | Change packing to one with more suitable lubricants or fit lantern ring with lubricant feed. Flush abrasives |

**Figure 6.10 Soft packing rings after use.** Left: *normal appearance*; top: *scored ring due to rotation of the ring in its housing;* right: *extruded ring due to excessive clearance between housing and shaft*

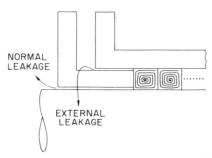

**Figure 6.11 Packed gland showing abnormal leakage outside the gland follower**

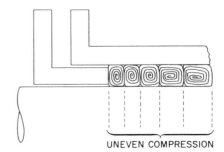

**Figure 6.12 Packed gland showing uneven compression due to incorrect installation**

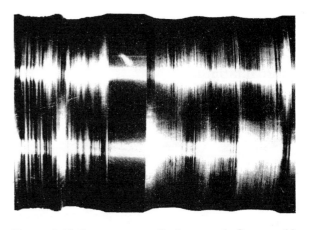

**Figure 6.13 Severe wear of a bronze shaft caused by a soft packing with inadequate boundary lubricant**

# Wire rope failures

A wire rope is said to have failed when the condition of either the wire strands, core or termination has deteriorated to an unacceptable extent. Each application has to be considered individually in terms of the degree of degradation allowable; certain applications may allow for a greater degree of deterioration than others.

Complete wire rope failures rarely occur. The more common modes of failure/deterioration are described below.

## DETERIORATION

### Mechanical damage

#### Characteristics

Damage to exposed wires or complete strands, often associated with gross plastic deformation of the steel material. Damage may be localised or distributed along the length of the rope.

Inspection by visual means only.

#### Causes

There are many potential causes of mechanical damage, such as:

- rubbing against a static structure whilst under load
- impact or collision by a heavy object
- misuse or bad handling practices

### External wear

#### Characteristics

Flattened areas formed on outer wires. Wear may be distributed over the entire surface or concentrated in narrow axial zones. Severe loss of worn wires under direct tension. Choice of rope construction can be significant in increasing wear resistance (e.g. Lang's lay ropes are usually superior to ordinary lay ropes).

Assess condition visually and also by measuring the reduction in rope diameter.

#### Causes

Abrasive wear between rope and pulleys, or between successive rope layers in multi-coiled applications, particularly in dirty or contaminated conditions (e.g. mining). Small oscillations, as a result of vibration, can cause localised wear at pulley positions.

Regular rope lubrication (dressings) can help to reduce this type of wear.

### External fatigue

#### Characteristics

Transverse fractures of individual wires which may subsequently become worn. Fatigue failures of individual wires occur at the position of maximum rope diameter ('crown' fractures).

Condition is assessed by counting the number of broken wires over a given length of rope (e.g. one lay length, 10 diameters, 1 metre).

#### Causes

Fatigue failures of wires is caused by cyclic stresses induced by bending, often superimposed on the direct stress under tension. Tight bend radii on pulleys increases the stresses and hence the risk of fatigue. Localised Hertzian stresses resulting from ropes operating in oversize or undersize grooves can also promote premature fatigue failures.

## Internal damage

### Characteristics

Wear of internal wires generates debris which when oxidised may give the rope a rusty (or 'rouged') appearance, particularly noticeable in the valleys between strands.

Actual internal condition can only be inspected directly by unwinding the rope using clamps while under no load.

As well as a visual assessment of condition, a reduction in rope diameter can give an indication of rope deterioration.

### Causes

Movement between strands within the rope due to bending or varying tension causes wear to the strand cross-over points (nicks). Failure at these positions due to fatigue or direct stress leads to fracture of individual wires. Gradual loss of lubricant in fibre core ropes accelerates this type of damage.

Regular application of rope dressings minimises the risk of this type of damage.

## Corrosion

### Characteristics

Degradation of steel wires evenly distributed over all exposed surfaces. Ropes constructed with galvanised wires can be used where there is a risk of severe corrosion.

### Causes

Chemical attack of steel surface by corrosive environment e.g. seawater.

Regular application of rope dressings can be beneficial in protecting exposed surfaces.

## Deterioration at rope terminations

### Characteristics

Failure of wires in the region adjacent to the fitting. Under severe loading conditions, the fitting may also sustain damage.

### Causes

Damage to the termination fitting or to the rope adjacent to the fitting can be caused by localised stresses resulting from sideways loads on the rope.

Overloading or shock loads can result in damage in the region of the termination.

Poor assembly techniques (e.g. incorrect mounting of termination fitting) can give rise to premature deterioration at the rope termination.

All photographs courtesy of Bridon Ropes Ltd., Doncaster

# Wire rope failures

## INSPECTION

To ensure safety and reliability of equipment using wire ropes, the condition of the ropes needs to be regularly assessed. High standards of maintenance generally result in increased rope lives, particularly where corrosion or fatigue are the main causes of deterioration.

The frequency of inspections may be determined by either the manufacturer's recommendations, or based on experience of the rate of rope deterioration for the equipment and the results from previous inspections. In situations where the usage is variable, this may be taken into consideration also.

Inspection of rope condition should address the following items:
- mechanical damage or rope distortions
- external wear
- internal wear and core condition
- broken wires (external and internal)
- corrosion
- rope terminations
- degree of lubrication
- equality of rope tension in multiple-rope installations
- condition of pulleys and sheaves

During inspection, particular attention should be paid to the following areas:
- point of attachment to the structure or drums
- the portions of the rope at the entry and exit positions on pulleys and sheaves
- lengths of rope subject to reverse or multiple bends

In order to inspect the internal condition of wire ropes, special tools may be required.

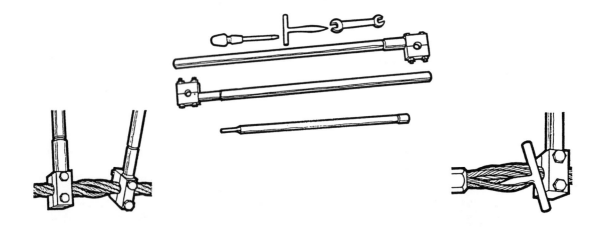

*Figure 7.1 Special tools for internal examination of wire rope*

## MAINTENANCE

Maintenance of wire ropes is largely confined to the application of rope dressings, general cleaning, and the removal of occasional broken wires.

Wire rope dressings are usually based on mineral oils, and may contain anti-wear additives, corrosion inhibiting agents or tackiness additives. Solvents may be used as part of the overall formulation in order to improve the penetrability of the dressing into the core of the rope. Advice from rope manufacturers should be sought in order to ensure that selected dressings are compatible with the lubricant used during manufacture.

The frequency of rope lubrication depends on the rate of rope deterioration identified by regular inspection. Dressings should be applied at regular intervals and certainly before there are signs of corrosion or dryness.

Dressings can be applied by brushing, spraying, dripfeed, or by automatic applicators. For best results, the dressing should be applied at a position where the rope strands are opened up such as when the rope passes over a pulley.

When necessary and practicable ropes can be cleaned using a wire brush in order to remove any particles such as dirt, sand or grit.

Occasional broken wires should be removed by using a pair of pliers to bend the wire end backwards and forwards until it breaks at the strand cross-over point.

## REPLACEMENT CRITERIA

Although the assessment of rope condition is mainly qualitative, it is possible to quantify particular modes of deterioration and apply a criterion for replacement. In particular the following parameters can be quantified:
- the number of wire breaks over a given length
- the change in rope diameter

Guidance for the acceptable density of broken wires in six and eight strand ropes is given below.

### Table 7.1 Criterion for replacement based on the maximum number of distributed broken wires in six and eight strand ropes operating with metal sheaves

| Total number of wires in outer strands (including filler wires) | Example of rope construction | Number of visible broken wires necessitating discard in a wire rope operating with metal sheaves when measured over a length of 10× nominal rope diameter | |
|---|---|---|---|
| | | Factor of safety <5 | Factor of safety >5 |
| Less than 50 | 6 × 7 (6/1) | 2 | 4 |
| 51–120 | 6 × 19 (9/9/1) | 3–5 | 6–10 |
| 121–160 | 6 × 19F (12/6+6F/1) | 6–7 | 12–14 |
| 161–220 | 8 × 19F (12/6+6F/1) | 8–10 | 16–20 |
| 221–260 | 6 × 37 (18/12/6/1) | 11–12 | 22–24 |

Rope manufacturers should be consulted regarding other types of rope construction.

Guidance for the allowable change in rope diameter is given below.

### Table 7.2 Criterion for replacement based on the change in diameter of a wire rope

| Rope construction | Replacement criteria related to rope diameter* |
|---|---|
| Six and eight strand ropes | Replacement necessary when the rope diameter is reduced to 90% of the nominal diameter at any position. |
| Multi-strand ropes | A more detailed rope examination is necessary when the rope diameter is either reduced to 97%, or has increased to 105%, of the nominal diameter. |

*The diameter of the rope is measured across the tips of the strands (i.e. the maximum rope diameter).

Some of the more common brake and clutch troubles are pictorially presented in subsequent sections; although these faults can affect performance and shorten the life of the components, only in exceptional circumstances do they result in complete failure.

## BRAKING TROUBLES

### Metal surface

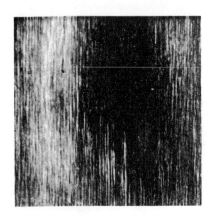

**Heat Spotting**

*Characteristics*

Small isolated discoloured regions on the friction surface. Often cracks are formed in these regions owing to structural changes in the metal, and may penetrate into the component.

*Causes*

Friction material not sufficiently conformable to the metal member; or latter is distorted so that contact occurs only at small heavily loaded areas.

**Crazing**

*Characteristics*

Randomly orientated cracks on the rubbing surface of a mating component, with main cracks approximately perpendicular to the direction of rubbing. These can cause severe lining wear.

*Causes*

Overheating and repeated stress-cycling from compression to tension of the metal component as it is continually heated and cooled.

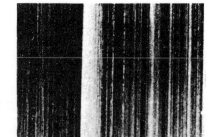

**Scoring**

*Characteristics*

Scratches on the rubbing path in the line of movement.

*Causes*

Metal too soft for the friction material; abrasive debris embedded in the lining material.

### Friction material surface

**Heat Spotting**

*Characteristics*

Heavy gouging caused by hard proud spots on drum resulting in high localised work rates giving rise to rapid lining wear.

*Causes*

Material rubbing against a heat-spotted metal member.

**Crazing**

*Characteristics*

Randomly orientated cracks on the friction material, resulting in a high rate of wear.

*Causes*

Overheating of the braking surface from overloading or by the brakes dragging.

**Scoring**

*Characteristics*

Grooves formed on the friction material in the line of movement, resulting in a reduction of life.

*Causes*

As for metal surface or using new friction material against metal member which needs regrinding.

## Fade

*Characteristics*

Material degrades at the friction surface, resulting in a decrease in $\mu$ and a loss in performance, which may recover.

*Causes*

Overheating caused by excessive braking, or by brakes dragging.

## Metal Pick-up

*Characteristics*

Metal plucked from the mating member and embedded in the lining.

*Causes*

Unsuitable combination of materials.

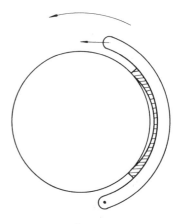

## Grab

*Characteristics*

Linings contacting at ends only ('heel and toe' contact) giving high servo effect and erratic performance. The brake is often noisy.

*Causes*

Incorrect radiusing of lining.

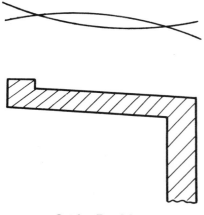

## Strip Braking

*Characteristics*

Braking over a small strip of the rubbing path giving localised heating and preferential wear at these areas.

*Causes*

Distortion of the brake path making it concave or convex to the lining, or by a drum bell mouthing.

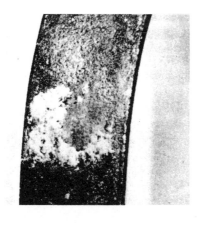

## Neglect

*Characteristics*

Material completely worn off the shoe giving a reduced performance and producing severe scoring or damage to the mating component, and is very dangerous.

*Causes*

Failure to provide any maintenance.

## Misalignment

*Characteristics*

Excessive grooving and wear at preferential areas of the lining surface, often resulting in damage to the metal member.

*Causes*

Slovenly workmanship in not fitting the lining correctly to the shoe platform, or fitting a twisted shoe or band.

## CLUTCH TROUBLES

As with brakes, heat spotting, crazing and scoring can occur with clutches; other clutch troubles are shown below.

### *Dishing

*Characteristics*

Clutch plates distorted into a conical shape. The plates then continually drag when the clutch is disengaged, and overheating occurs resulting in thermal damage and failure. More likely in multi-disc clutches.

*Causes*

Lack of conformability. The temperature of the outer region of the plate is higher than the inner region. On cooling the outside diameter shrinks and the inner area is forced outwards in an axial direction causing dishing.

### *Waviness or Buckling

*Characteristics*

Clutch plates become buckled into a wavy pattern. Preferential heating then occurs giving rise to thermal damage and failure. More likely in multi-disc clutches.

*Causes*

Lack of conformability. The inner area is hotter than the outer area and on cooling the inner diameter contracts and compressive stresses occur in the outer area giving rise to buckling.

### *Band Crushing

*Characteristics*

Loss of friction material at the ends of a band in a band clutch. Usually results in grooving and excessive wear of the opposing member.

*Causes*

Crushing and excessive wear of the friction material owing to the high loads developed at the ends of a band of a positive servo band clutch.

### *Bond Failure

*Characteristics*

Material parting at the bond to the core plate causing loss of performance and damage to components.

*Causes*

Poor bonding or overheating, the high temperatures affecting bonding agent.

### Material Transfer

*Characteristics*

Friction material adhering to opposing plate, often giving rise to excessive wear.

*Causes*

Overheating and unsuitable friction material.

### Burst Failure

*Characteristics*

Material splitting and removed from the spinner plate.

*Causes*

High stresses on a facing when continually working at high rates of energy dissipation, and high speeds.

*These refer to oil immersed applications.

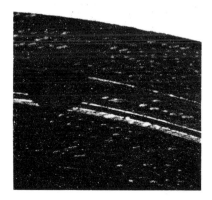

### Grooving

*Characteristics*

Grooving of the facing material on the line of movement.

*Causes*

Material transfer to opposing plate.

### Reduced Performance

*Characteristics*

Decrease in coefficient of friction giving a permanent loss in performance in a dry clutch.

*Causes*

Excess oil or grease on friction material or on the opposing surface.

### Distortion

*Characteristics*

Facings out of flatness after high operating temperatures giving rise to erratic clutch engagement.

*Causes*

Unsuitable friction material.

## GENERAL NOTES

The action required to prevent these failures recurring is usually obvious when the causes, as listed in this section, are known.

Other difficulties can be experienced unless the correct choice of friction material is made for the operating conditions.

If the lining fitted has too low a coefficient of friction the friction device will suffer loss of effectiveness. Oil and grease deposited on dry linings and facings can have an even more marked reduction in performance by a factor of up to 3. If the $\mu$ is too high or if a badly matched set of linings are fitted, the brake may grab or squeal.

The torque developed by the brake is also influenced by the way the linings are bedded so that linings should be initially ground to the radius of the drum to ensure contact is made as far as possible over their complete length.

If after fitting, the brake is noisy the lining should be checked for correct seating and the rivets checked for tightness. All bolts should be tightened and checks made that the alignment is correct, that all shoes have been correctly adjusted and the linings are as fully bedded as possible. Similarly, a clutch can behave erratically or judder if the mechanism is not correctly aligned.

## BASIC MECHANISMS

Fretting occurs where two contacting surfaces, often nominally at rest, undergo minute oscillatory tangential relative motion, which is known as 'slip'. It may manifest itself by debris oozing from the contact, particularly if the contact is lubricated with oil.

    Colour of debris: red on iron and steel, black on aluminium and its alloys.

On inspection the fretted surfaces show shallow pits filled and surrounded with debris. Where the debris can escape from the contact, loss of fit may eventually result. If the debris is trapped, seizure can occur which is serious where the contact has to move occasionally, e.g. a machine governor.

The movement may be caused by vibration, or very often it results from one of the contacting members undergoing cyclic stressing. In this case fatigue cracks may be observed in the fretted area. Fatigue cracks generated by fretting start at an oblique angle to the surface. When they pass out of the influence of the fretting they usually continue to propagate straight across the component. This means that where the component breaks, there is a small tongue of metal on one of the fracture surfaces corresponding to the growth of the initial part of the crack.

Fretting can reduce the fatigue strength by 70–80%. It reaches a maximum at an amplitude of slip of about 8 $\mu$m. At higher amplitudes of slip the reduction is less as the amount of material abraded away increases.

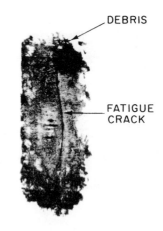

FRETTING SCAR

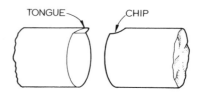

**Figure 9.1 A typical fatigue fracture initiated by fretting**

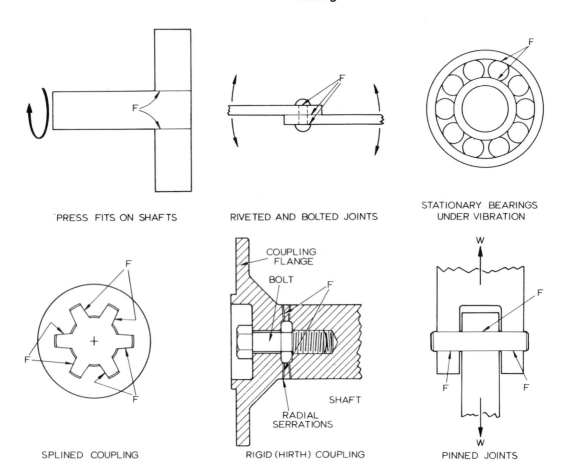

PRESS FITS ON SHAFTS      RIVETED AND BOLTED JOINTS      STATIONARY BEARINGS UNDER VIBRATION

SPLINED COUPLING      RIGID (HIRTH) COUPLING      PINNED JOINTS

**Figure 9.2 Typical situations in which fretting occurs. Fretting sites are at points F.**

## Detailed mechanisms

Rupture of oxide films results in formation of local welds which are subjected to high strain fatigue. This results in the growth of fatigue cracks oblique to the surface. If they run together a loose particle is formed. One of the fatigue cracks may continue to propagate and lead to failure. Oxidation of the metallic particles forms hard oxide debris, i.e. $Fe_2O_3$ on steel, $Al_2O_3$ on aluminium. Spreading of this oxide debris causes further damage by abrasion. If the debris is compacted on the surfaces the damage rate becomes low.

Where the slip is forced, fretting wear damage increases roughly linearly with normal load, amplitude of slip, and number of cycles. Damage rate on mild steel—approx. 0.1 mg per $10^6$ cycles, per $MN/m^2$ normal load, per $\mu$m amplitude of slip. Increasing the pressure can, in some instances, reduce or prevent slip and hence reduce fretting damage.

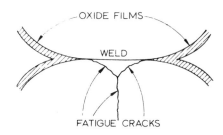

**Figure 9.3 Oxide film rupture and the development of fatigue cracks**

## PREVENTION

### Design

(a) elimination of stress concentrations which cause slip
(b) separating surfaces where fretting is occurring
(c) increasing pressure by reducing area of contact

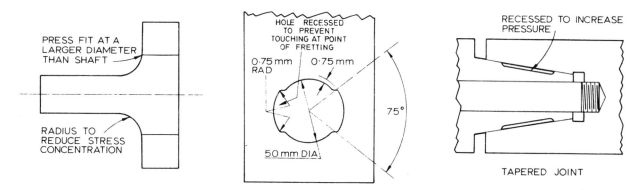

**Figure 9.4 Design changes to reduce the risk of fretting**

## Lubrication

Where the contact can be continuously fed with oil, the lubricant prevents access of oxygen which is advantageous in reducing the damage. Oxygen diffusion decreases as the viscosity increases. Therefore as high a viscosity as is compatible with adequate feeding is desirable. The flow of lubricant also carries away any debris which may be formed. In other situations greases must be used. Shear-susceptible greases with a worked penetration of 320 are recommended. E.P. additives and $MoS_2$ appear to have little further beneficial effect, but anti-oxidants may be of value. Baked-on $MoS_2$ films are initially effective but gradually wear away.

## Non-metallic coatings

Phosphate and sulphidised coatings on steel and anodised coatings on aluminium prevent metal-to-metal contact. Their performance may be improved by impregnating them with lubricants, particularly oil-in-water emulsions.

## Metallic coatings

Electrodeposited coatings of soft metals, e.g. Cu, Ag, Sn or In or sprayed coatings of Al allow the relative movement to be taken up within the coating. Chromium plating is generally not recommended.

## Non-metallic inserts

Inserts of rubber, or PTFE can sometimes be used to separate the surfaces and take up the relative movement.

## Choice of metal combinations

Unlike metals in contact are recommended—preferably a soft metal with low work hardenability and low recrystallisation temperature (such as Cu) in contact with a hard surface, e.g. carburised steel.

# 10        Maintenance methods

The purpose of maintenance is to preserve plant and machinery in a condition in which it can operate, and can do so safely and economically.

### Table 10.1  Situations requiring maintenance action

| Situation | Basis of decision | Decision mechanism |
|---|---|---|
| Loss of operating efficiency | Recognition of the problem associated with a study of the costs | The economic loss arising from the reduced efficiency is greater than the effective cost of the repair. |
| Loss of function | Safety risk | This must be avoided by prior maintenance action. There must therefore be a method of predicting the approach of a failure condition. |
| | Downtime cost<br>Repair cost | The maintenance method that is selected needs to be chosen to keep total costs to a minimum, while maximising the profit and market opportunities.<br><br>The ongoing cost of maintenance versus total replacement also needs to be considered. |

It is the components of plant and machinery which fail individually, and can lead to the loss of function of the whole unit or system. Maintenance activity needs therefore to be concentrated on those components that are critical.

### Table 10.2  Factors in the selection of critical components

| Important factor | Typical examples | Guide to selection |
|---|---|---|
| Likelihood of failure | Components subject to:<br>Occasional overload<br>Fatigue loading<br>Wear<br>Corrosion or other environmental effects | Can be identified by a review of the design of the system or by an analysis of relevant operating experience.<br>Components which are more likely to fail should be selected for maintenance attention. |
| Effect of the component failure on the system | Components which, when they fail, cause a failure of the whole system, possibly with some consequential damage | These components need to be selected for special attention.<br>They must be selected if there are safety implications. |
| | Components which, when they fail, may allow other components to be overloaded, and then cause failure of the system | An analysis of the effects of the failure needs to be carried out to detect any safety implications. |
| | Components which, when they fail, only produce a reduction in the performance of the system | The likely economic effects of the failure need to be analysed to decide whether these components merit special attention. |
| The time required to replace a failed component or to rectify the effect of its failure | Components that can be replaced quickly such as bulbs, fuses and some printed circuit boards, or components whose function can be taken over immediately by a standby system | If there are no safety implications, no particular maintenance action is required other than component replacement or repair after failure |
| | Components which take a long time to replace or repair | Their condition needs to be monitored and their maintenance planned in advance |

## Table 10.3 Maintenance methods which can be used

| Maintenance method | Advantages | Disadvantages |
| --- | --- | --- |
| Allow the equipment to break down and then repair it | Can be the cheapest solution if the repair is easy and there is no safety risk or possibility of consequential damage | The cost of consequential damage can be high and there might be safety risks. The plant may be of a type which cannot be allowed to stop suddenly because of product solidification or deterioration, etc.<br><br>The failure may occur at an inconvenient time or, if the plant is mobile, at an inconvenient place. The necessary staff and replacement components may not then be available. |
| Preventive maintenance carried out at regular intervals | Reduces the risk of failure in service<br><br>Enables the work and availability of special tools and spare parts to be planned well in advance<br><br>Is particularly appropriate for components which need changing because of capacity absorption such as filter elements | Some failures will continue to occur in service because components do not fail at regular intervals as shown in Figure 10.2. The failures can be unexpected and inconvenient.<br><br>During the overhaul many components in good condition will be stripped and inspected unnecessarily. Mistakes can be made on re-assembly so that the plant can end up in a worse condition after the overhaul.<br><br>The overhaul process can take a considerable time resulting in a major loss of profitable production. |
| Opportunistic maintenance carried out when the plant or equipment happens to become available | The maintenance can be carried out with no loss of effective operating time. | Staff and spare parts have to be kept available which may cause increased costs.<br><br>Is only practical in combination with condition based maintenance or some preventive maintenance, because otherwise any work which needs to be done is not known about.<br><br>Breakdowns are still likely to occur in service. |
| Condition-based maintenance | Enables the plant to be left in service until a failure is about to occur. It can then be withdrawn from service in a planned manner and repaired at minimum cost.<br><br>Consequential damage from a failure can be avoided. | Can only detect failures which show a progression to failure, which enables the incipient failure to be detected. Components with a sudden failure mode cannot provide the required advanced detection.<br><br>Failures of components due to an unexpected random overload in service cannot be avoided, even by this method. |
| Maintenance/replacement based on expected component life | Is the only safe method usable for components which have a sudden failure mode | Requires monitoring of the operating conditions of the component so that its expected life can be estimated.<br><br>If all failures are to be avoided many components will need to be changed when individually they have a useful life still remaining, e.g. in Figure 10.2 the working life has to be kept below the range in which failures may occur. |

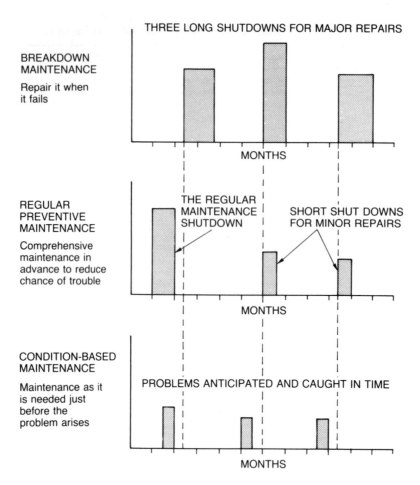

**Figure 10.1 The results of maintaining the same industrial plant in different ways**
*The height of the bars indicates the amount of maintenance effort required*

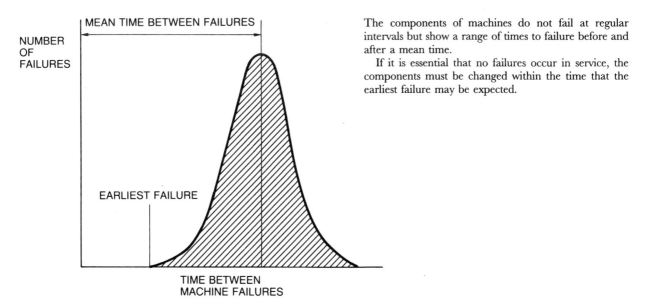

The components of machines do not fail at regular intervals but show a range of times to failure before and after a mean time.

If it is essential that no failures occur in service, the components must be changed within the time that the earliest failure may be expected.

**Figure 10.2 The distribution of the time to failure for a typical component**

## Table 10.4 Feedback from maintenance to equipment design

| Opportunity | Remarks |
|---|---|
| The maintenance activity provides an opportunity to record and analyse the problems and failures which occur during the service life of equipment. In particular, records need to be kept of the operating time to failure, the failure mode and any particular operating conditions which may have contributed to the failure. | This is a unique source of information which is of very high value to the process of design and development of improved equipment.<br><br>It is a profit centre of the maintenance activity. |
| An intelligent review of maintenance operations can enable repair times to be determined for various components.<br><br>The nature and cause of accessibility problems can be determined together with the definition of the design principles which need to be used to improve it. | This is useful for maintenance planning but it is particularly important as a method of getting new equipment that is designed to make maintenance more simple and of reduced cost. |
| The maintenance activity can be reviewed to identify opportunities in which modular design can be used to reduce downtime caused by in service faults. It can also often simplify the process of equipment manufacture. | If sub-systems can be designed to be self-contained and readily removable, this can enable the sub-system to be exchanged when a fault occurs.<br><br>The actual problem can then be corrected off-line at a comprehensive test facility and the sub-system returned for further use as an exchange unit. |

### Table 10.5 Maintenance management

| Technique | Description | Remarks |
|---|---|---|
| A separate maintenance department | Called in by the operating departments when something goes wrong | Inherently operates on a breakdown basis. Has little chance to contribute to a major uplift in plant performance. An outdated technique. |
| Terotechnology | Viewing plant and equipment in terms of their total life cost and thus placing more emphasis on cost effective maintainability as distinct from first cost alone | A level of management philosophy which prepares the way for efficient maintenance systems |
| Computerised maintenance information | Systems covering: Plant inventory and data Operating history Spares usage and storage On-going condition data Repair work records Failure analysis and reliability data Cost records | Provides the basis for an effective management system that is able to concentrate on problem areas, plan ahead, reduce costs and contribute to the specification of improved future plant |
| Total productive maintenance | A technique developed in Japan in which the production operators carry out the first line of problem detection and maintenance. There is intentionally no clear interface between the production and maintenance departments. | An effective way of involving all the people in a company and achieving recognition of the importance of process plant availability. It can reduce in-service failures by 25%. |
| Reliability centred maintenance | A technique developed in the airlines. A system which looks at component functions, failure modes and effects in order to concentrate on key issues. Safety in operation is paramount followed closely by the control of downtime, quality and customer service. | An effective means of concentrating maintenance effort cost effectively, with an ability to identify problems and any need for design improvements |
| Mobile equipment maintenance | The basic objective is to ensure that the equipment does not break down away from its base. Also, when visiting its base, any essential work that is needed should be carried out. | It is particularly important to monitor the condition of such plant and its expected component lives. It is sometimes necessary to carry out some maintenance work before it becomes essential in order to match the time of access to the equipment, and the overhaul department's workload. |

# Condition monitoring

Condition monitoring is a technique used to monitor the condition of equipment in order to give an advanced warning of failure. It is an essential component of condition-based maintenance in which equipment is maintained on the basis of its condition.

## MONITORING METHODS

The basic principle of condition monitoring is to select a physical measurement which indicates that deterioration is occurring, and then to take readings at regular intervals. Any upward trend can then be detected and taken as an indication that a problem exists. This is illustrated in Figure 11.1 which shows a typical trend curve and the way in which this provides an alert that an incipient failure is approaching. It also gives a lead time in which to plan and implement a repair.

Since failures occur to individual components, the monitoring measurements need to focus on the particular failure modes of the critical components.

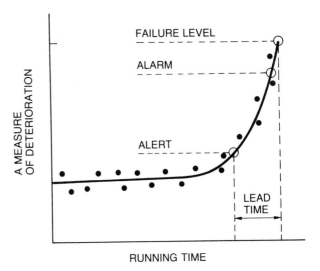

**Figure 11.1 The principle of condition monitoring measurements which give an indication of the deterioration of the equipment**

## Table 11.1 Monitoring methods and the components for which they are suitable

| Method | Principle | Application examples |
|---|---|---|
| Wear debris monitoring | The collection and analysis of wear debris derived from component surfaces, and carried away in the lubricating oil | Components such as bearings or other rubbing parts which wear, or suffer from surface pitting due to fatigue |
| Vibration monitoring | The detection of faults in moving components, from the change in the dynamic forces which they generate, and which affect vibration levels at externally accessible points | Rotating components such as gears and high speed rotors in turbines and pumps |
| Performance monitoring | Checking that the machine components and the complete machine system are performing their intended functions | The temperature of a bearing indicates whether it is operating with low friction. The pressure and flow rate of a pump indicate whether its internal components are in good condition. |

The monitoring measurements give an indication of the existence of a problem as shown in Figure 11.1. More detailed analysis can indicate the nature of the problem so that rectification action can be planned. Other sections of this handbook give more details about these methods of monitoring.

In wear debris monitoring, the amount of the debris and its rate of generation indicate when there is a problem. The material and shape of the debris particles can indicate the source and the failure mechanism.

The overall level of a vibration measurement can indicate the existence of a problem. The form and frequency of the vibration signal can indicate where the problem is occurring and what it is likely to be.

# Condition monitoring

## Introducing condition monitoring

If an organisation has been operating with breakdown maintenance or regular planned maintenance, a change over to condition-based maintenance can result in major improvements in plant availability and in reduced costs. There are, however, up front costs for organisation and training and for the purchase of appropriate instrumentation. There are operational circumstances which can favour or retard the potential for the introduction of condition-based maintenance.

### Table 11.2 Factors which can assist the introduction of condition-based maintenance

| Factor | Mechanism of action |
| --- | --- |
| Where a safety risk is particularly likely to arise from the breakdown of machinery | Typical examples are plant handling dangerous materials, and machines for the transport of people. |
| Where accurate advanced planning of maintenance is essential | Typical examples are equipment situated in a remote place which is visited only occasionally for maintenance, and mobile equipment which makes only occasional visits to its base. |
| Where plant or equipment is of recent design, and may have some residual development problems | Condition monitoring enables faults to be detected early while damage is still slight, thus providing useful evidence to guide design improvements. It also improves the negotiating position with the plant manufacturer. |
| Where relatively insensitive operators use expensive equipment whose breakdown may result in serious damage | Condition monitoring enables a fault to be detected in sufficient time for an instruction to be issued for the withdrawal of the equipment before expensive damage is done. |
| Where the manufacturer can offer a condition monitoring service to several users of his equipment | The cost to each user can be reduced in this way, and the manufacturer gets a useful feed-back to guide his product design and development. |
| Where instruments or other equipment required for condition monitoring can be used, or is already being used, for another purpose | Other applications of the instruments or equipment may be process control or some servicing activity such as rotor balancing. |

### Table 11.3 Factors which can retard the introduction of condition-based maintenance

| Factor | Mechanism of action |
| --- | --- |
| Where an industry is operating at a low level of activity, or operates seasonally, so that plant and machinery is often idle | If the plant is only operating part of the time, there is generally plenty of opportunity for inspection and maintenance during idle periods. |
| Where there is too small a number of similar machines or components being monitored by one engineer or group of engineers to enable sufficient experience to be built up for the effective interpretation of readings and for correct decisions on their significance | To gain experience in a reasonable time, the minimum number of machines tends to vary between 4 and 10 depending on the type of machine or component. The problem may be overcome by pooling monitoring services with other companies, or by involving machine manufacturers or external monitoring services. |
| Where skilled operators have close physical contact with their machines, and can use their own senses for subjective monitoring | Machine tools and ships can be examples of this situation, but any trends towards the use of less skilled operators or supervisory engineers, favours the application of condition monitoring. |

## Table 11.4 A procedure for setting up a plant condition monitoring activity

| Activity | Remarks |
|---|---|
| 1. Check that the plant is large enough to justify having its own internal system. | If the total plant value is less than £2M it may be worth sub-contracting the activity. |
| 2. Consider the cost of setting up. | For most plant, a setting up cost of 1% of the plant value can be justified. If there is a major safety risk, up to 5% of the plant value may be appropriate. |
| 3. Select the machines in the plant that should be monitored. | The important machines for monitoring will tend to be those which:<br>(a) Are in continuous operation.<br>(b) Are involved in single stream processes.<br>(c) Have minimum parallel or stand-by capacity.<br>(d) Have the minimum product storage capacity on either side of them.<br>(e) Handle dangerous or toxic materials.<br>(f) Operate to particularly high pressures or speeds. |
| 4. Select the components of the critical machines on which the monitoring needs to be focussed. | The important components will be those where:<br>(a) A failure is possible.<br>(b) The consequences of the failure are serious in terms of safety or machine operation.<br>(c) If a failure is allowed to occur the time required for a repair is likely to be long. |
| 5. Choose the monitoring method or methods to be used. | List the possible techniques for each critical component and try to settle for two or at the most three techniques for use on the plant. |

## Table 11.5 Problems which can arise

| Problem | Solution |
|---|---|
| Regular measurements need to be taken, often for months or years before a critical situation arises. The operators can therefore get bored. | The management need to keep the staff motivated by stressing the importance of their work.<br><br>The use of portable electronic data collectors partially automates the collection process, provides a convenient interface with a computer for data analysis, and can also monitor the tour of duty of the operators. |
| One of the measurements indicates that an alert situation has arisen and a decision has to be made on whether to shut down the plant and incur high costs from loss of use, or whether it is a false alarm. | To avoid this situation install at least two physically different systems for monitoring really critical components. e.g. measure bearing temperature and vibration.<br><br>In any event always recheck deviant readings and re-examine past trends. |
| The operators take a long time to acquire the necessary experience in detection and diagnosis, and can create false alarms. | Start taking the measurements while still operating a planned regular maintenance procedure.<br><br>Take many measurements just prior to shut down and then check the components to see whether the diagnosis was correct. |

### Table 11.6 The benefits that can arise from the use of condition monitoring

| Benefit | Mechanism |
| --- | --- |
| 1. Increased plant availability resulting in greater output from the capital invested.<br><br>2. Reduced maintenance costs. | Machine running time can be increased by maximising the time between overhauls. Overhaul time can be reduced because the nature of the problem is known, and the spares and men can be ready. Consequential damage can be reduced or eliminated. |
| 3. Improved operator and passenger safety. | The lead time given by condition monitoring enables machines to be stopped before they reach a critical condition, especially if instant shut-down is not permitted. |
| 4. More efficient plant operation, and more consistent quality, obtained by matching the rate of output to the plant condition. | The operating load and speed on some machines can be varied to obtain a better compromise between output, and operating life to the next overhaul. |
| 5. More effective negotiations with plant manufacturers or repairers, backed up by systematic measurements of plant condition. | Measurements of plant when new, at the end of the guarantee period, and after overhaul, give useful comparative values. |
| 6. Better customer relations following from the avoidance of inconvenient breakdowns which would otherwise have occurred. | The lead time given by condition monitoring enables such breakdowns to be avoided. |
| 7. The opportunity to specify and design better plant in the future. | The recorded experience of the operation of the present machinery is used for this purpose. |

## Table 12.1 Maximum contact temperatures for typical tribological components

| Component | Maximum temperature | Reason for limitation |
|---|---|---|
| White metal bearing | 200°C at 1.5 MN/m² to 130°C at 7 MN/m² | Failure by incipient melting at low loading (1.5 MN/m²); by plastic deformation at high loading (7 MN/m²) |
| Rolling bearing | 125°C | Normal tempering temperature (special bearings are available for higher temperature operation) |
| Steel gear | 150–250°C | Scuffing; the temperature at which scuffing occurs is a function of both the lubricant and the steel and cannot be defined more closely |

The temperatures in Table 12.1 are indicative of design limits. In practice it may be difficult to measure the contact temperature. Table 12.2 indicates practical methods of measuring temperatures and the limits that can be accepted.

## Table 12.2 Temperature as an indication of component failure

| Component | Method of temperature measurement | Comments | Action limits [1] [4] |
|---|---|---|---|
| White metal bearing | Thermocouple in contact with back of white metal in thrust pad or at load line in journal bearings [5] | Extremely sensitive, giving immediate response to changes in load. Failure is indicated by rapid temperature rise | Alarm at rise of 10°C above normal running temperature. Trip at rise of 20°C |
| | Thermometer/thermocouple in oil bleed from bearing (viz. through hole drilled in bearing land) | Reasonably sensitive, may be preferable for journal bearings where there is difficulty in fitting a thermocouple into the back of the bearing in the loaded area | Alarm at rise of 10°C above normal running temperature. Trip at rise of 20°C |
| | Thermometer in bearing pocket or in drain oil | Relatively insensitive as majority of heat is carried away in oil that passes through bearing contact and this is rapidly cooled by excess oil that is fed to bearing. Can be useful in commissioning or checking replacements | Normal design 60°C Acceptable limit 80°C |
| Rolling bearing | Thermocouple or thermometer in contact with outer race (inner race rotating) | Two failure mechanisms cause temperature rise [2] | |
| | | (a) breakdown of lubrication | Slow rise of temperature from steady value is indicative of deterioration of lubrication: Alarm at 10°C rise. Acceptable limit 100°C [3] |
| | | (b) loss of internal clearance | Failure occurs so rapidly that there is insufficient time for warning of failure to be obtained from temperature indication of outer race |
| | Thermometer in oil | | Acceptable limit 100°C |
| Gears | Thermometer in oil | | Acceptable limit 80°C above ambient |
| Metallic packing | Contact thermometer on rod | | Acceptable limit 80°C |

(1) Temperature rise above normal value is more useful as an indication of trouble than the absolute value. The more the running value is below the acceptable limit the greater the margin of safety.
(2) Failure by fatigue or wear of raceways does not give temperature rise. They may be detected by an increase in noise level.
(3) Temperature in grease-packed bearing will rise to peak value until grease clears into housing and then fall to normal running value. Peak value may be 10–20°C above normal and attainment of equilibrium may take up to six hours. With bearing with grease relief valve a similar cycle will occur on each re-lubrication.
(4) Running-in. Higher than normal temperatures may occur during the initial running. Equilibrium temperatures can be expected after about twenty-four hours. The acceptable limits given should not be exceeded; if the limit is reached the machine should be stopped and and allowed to cool before proceeding with the run-in.
(5) Care must be taken to avoid deforming the bearing surface as this will result in a falsely high reading.

# 13             Vibration analysis

## PRINCIPLES

Vibration analysis uses vibration measurements taken at an accessible position on a machine, and analyses these measurements in order to infer the condition of moving components inside the machine.

### Table 13.1 The generation and transmission of vibration

| The signal | Mechanism | Examples |
|---|---|---|
| Generation of the signal | The mass centres of moving parts move during machine operation, generally in a cyclic manner. This gives rise to cyclic force variations. | Unbalanced shafts.<br>Bent shafts and resonant shafts<br>Rolling elements in rolling bearings moving unevenly.<br>Gear tooth meshing cycles.<br>Loose components.<br>Cyclic forces generated by fluid interactions. |
| Transmission of the signal | From the moving components via their supporting bearing components to the machine casing | Ideally, there should be a relatively rigid connecting path between the area where the vibration is transmitted internally to the machine casing, and the points on the outside of the machine where the measurement is taken. |
| Transmission problems | If the moving parts are very light and the machine casing is very heavy and rigid, the signal measured externally may be too small for accurate analysis and diagnosis. | High speeds rotors in high pressure machines with rigid barrel casings can have this problem.<br><br>A solution is to take a direct measurement of the cyclic movement of the shaft, relative to the casing at its supporting bearings. |

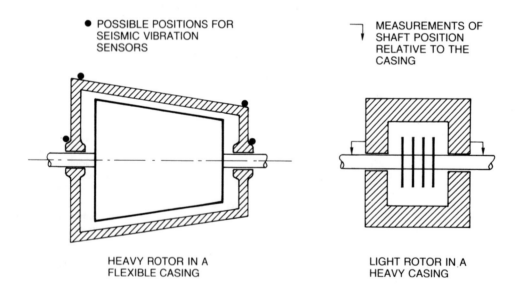

● POSSIBLE POSITIONS FOR SEISMIC VIBRATION SENSORS

MEASUREMENTS OF SHAFT POSITION RELATIVE TO THE CASING

HEAVY ROTOR IN A FLEXIBLE CASING

LIGHT ROTOR IN A HEAVY CASING

*Figure 13.1 Vibration measurements on machines*

## Table 13.2 Categories of vibration measurement

| Measurement | The principle behind the technique | Applications |
|---|---|---|
| Overall level of vibration (see subsequent section) | The general level of vibration over a wide frequency band. It determines the degree to which the machine may be running roughly.<br><br>It is a means of quantifying the technique of feeling a machine by hand. | All kinds of rotating machines but with particular application to higher speed machines<br><br>Not usually applicable to reciprocating machines |
| Spectral analysis of vibration (see subsequent section on vibration frequency monitoring) | The vibration signal is analysed to determine any frequencies where there is a substantial component of the vibration level. It is equivalent to scanning the frequency bands on a radio receiver to see if any station is transmitting. | From the value of the frequencies where there is a signal peak, the likely source of the vibration can be determined. Such a frequency might be the rotational speed of a particular shaft, or the tooth meshing frequency of a particular pair of gear wheels. |
| Discrete frequency monitoring | A method of monitoring a particular machine component by measuring the vibration level generated at the particular frequency which that component would be expected to generate | If a particular shaft in a machine is to be examined for any problems, the monitoring would be tuned to its rotational speed. |
| Shock pulse monitoring | Using a vibration probe, with a natural resonant frequency that is excited by the shocks generated in rolling element bearings, when they operate with fatigue pits in the surfaces of their races | The monitoring of rolling element bearings with a simple hand held instrument |
| Kurtosis measurement | This is a technique that looks at the 'spikyness' of a vibration signal, i.e. the number of sharp peaks as distinct from a smoother sinusoidal profile. | The monitoring of fatigue development in rolling bearings with a simple portable instrument, that is widely applicable to all types and sizes of bearing |
| Signal averaging (see subsequent section) | The accumulation over a few seconds of the parts of a cyclic vibration signal, which contain a particular frequency. Parts of the signal at other frequencies are averaged out.<br><br>By matching the particular frequency to, for example, the rotational speed of a particular machine component, the resulting diagram will show the characteristics of that component. | The monitoring of a gear by signal averaging, relative to its rotational speed, will show the cyclic action of each tooth. A tooth with a major crack could be detected by its increased flexibility. |
| Cepstrum analysis | If two vibration frequencies are superimposed in one signal, sideband frequencies are generated on either side of the higher frequency peak, with a spacing related to that of the lower frequency involved.<br><br>Cepstrum analysis looks at these sidebands in order to understand the underlying frequency patterns and their relative effects. | Interactions between the rotational frequency of bladed rotors and the blade passing frequency<br><br>Also between gear tooth meshing frequencies and gear rotational speeds |

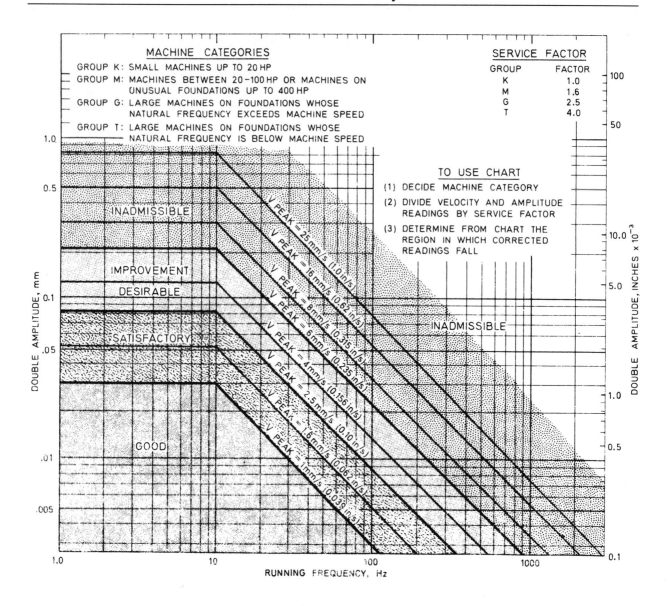

**Figure 13.2 Guidance on the levels of overall vibration of machines**

## OVERALL LEVEL MONITORING

This is the simplest method for the vibration monitoring of complete machines. It uses the cheapest and most compact equipment. It has the disadvantage however that it is relatively insensitive, compared with other methods, which focus more closely on to the individual components of a machine.

The overall vibration level can be presented as a peak to peak amplitude of vibration, as a peak velocity or as a peak acceleration. Over the speed range of common machines from 10 Hz to 1000 Hz vibration velocity is probably the most appropriate measure of vibration level. The vibration velocity combines displacement and frequency and is thus likely to relate to fatigue stresses.

The normal procedure is to measure the vertical, horizontal and axial vibration of a bearing housing or machine casing and take the largest value as being the most significant.

As in all condition monitoring methods, it is the trend in successive readings that is particularly significant. Figure 13.2, however, gives general guidance on acceptable overall vibration levels allowing for the size of a machine and the flexibility of its mounting arrangements.

For machine with light rotors in heavy casings, where it is more usual to make a direct measurement of shaft vibration displacement relative to the bearing housing, the maximum generally acceptable displacement is indicated in the following table.

### Table 13.3 Allowable vibrational displacements of shafts

| Ratio Vibration displacement / Diametral clearance | Speed rev/min |
|---|---|
| 0.5 | 300 |
| 0.25 | 3000 |
| 0.1 | 12000 |

## VIBRATION FREQUENCY MONITORING

The various components of a machine generate vibration at characteristic frequencies. If a vibration signal is analysed in terms of its frequency content, this can give guidance on its source, and therefore on the cause of any related problem. This spectral analysis is a useful technique for problem diagnosis and is often applied, when the overall level of vibration of a machine exceeds normal values.

In spectral analysis the vibration signal is converted into a graphical plot of signal strength against frequency as shown in Figure 13.3, in this case for a single reduction gearbox.

In Figure 13.3 there are three particular frequencies which contribute to most of the vibration signal and, as shown in Figure 13.4, they will usually correspond to the shaft speeds and gear tooth meshing frequencies.

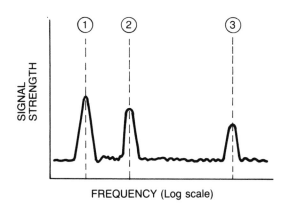

Figure 13.3 The spectral analysis of the vibration signal from a single reduction gearbox.

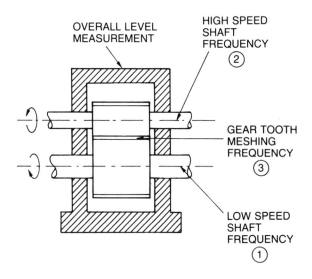

Figure 13.4 An example of the sources of discrete frequencies observable in a spectral analysis

## Discrete frequency monitoring

If it is required to monitor a particular critical component the measuring system can be turned to signals at its characteristic frequency in order to achieve the maximum sensitivity. This discrete frequency monitoring is particularly appropriate for use with portable data collectors, particularly if these can be preset to measure the critical frequencies at each measuring point. The recorded values can then be fed into a base computer for conversion into trends of the readings with the running time of the machine.

**Table 13.4 Typical discrete frequencies corresponding to various components and problems**

| Component/problem | Frequency | Characteristics |
|---|---|---|
| Unbalance in rotating parts | Shaft speed | Tends to increase with speed and when passing through a resonance such as a critical speed |
| Bent shaft | Shaft speed | Usually mainly axial vibration |
| Shaft misalignment | Shaft speed or $2 \times$ shaft speed | Often associated with high levels of axial vibration |
| Shaft rubs | Shaft speed and $2 \times$ shaft speed | Can excite higher resonant frequencies. May vary in level between runs. |
| Oil film whirl | 0.45 to $0.5 \times$ shaft speed | Only on machines with lubricated sleeve bearings |
| Gear tooth problems | Tooth meshing frequency | Generally also associated with noise |
| Reciprocating components | Running speed and $2 \times$ running speed. | Inherent in reciprocating machinery |
| Rolling element bearing fatigue damage | Shock pulses at high frequency | Caused by the rolling elements hitting the fatigue pits |
| Cavitation in fluid machines | High frequency similar to shock pulses | Can be mistaken for rolling element bearing problems |

## SIGNAL AVERAGING

If a rotating component carries a number of similar peripheral sub-units, such as the teeth on a gear wheel or the blades on a rotor which interact with a fluid, then signal averaging can be used as an additional monitoring method.

A probe is used to measure the vibrations being generated and the output from this is fed to a signal averaging circuit, which extracts the components of the signal which have a frequency base corresponding to the rotational speed of the rotating component which is to be monitored. This makes it possible to build up a diagram which shows how the vibration forces vary during one rotation of the component. Some typical diagrams of this kind are shown in Figure 13.5 which indicates the contribution to the vibration signal that is made by each tooth on a gear. An outline of the technique for doing this is shown in Figure 13.6.

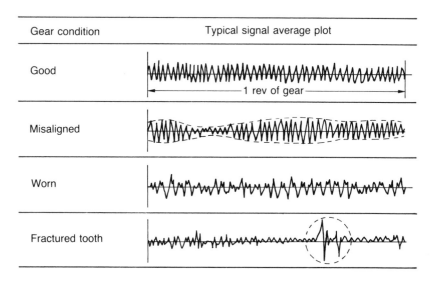

*Figure 13.5 Signal average plots used to monitor a gear and showing the contribution from each tooth*

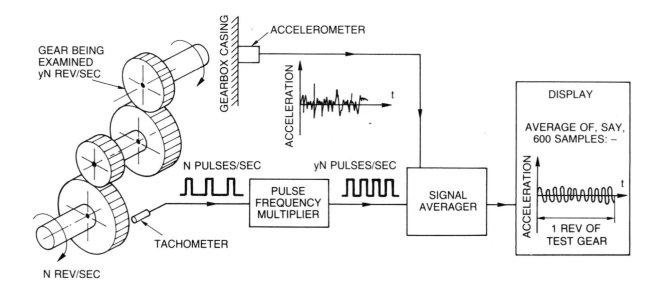

*Figure 13.6 A typical layout of a signal averaging system for monitoring a particular gear in a transmission system*

# 14            Wear debris analysis

In wear debris analysis machine lubricants are monitored for the presence of particles derived from the deterioration of machine components. The lubricant itself may also be analysed, to indicate its own condition and that of the machine.

## WEAR DEBRIS ANALYSIS

### Table 14.1 Wear debris monitoring methods

| Type | Mechanism of operation |
| --- | --- |
| **IN LINE** | |
| Monitoring the main flow of oil through the machine | Magnetic plugs or systems which draw ferro-magnetic particles from the oil flow for inspection, or on to an inductive sensor, that produces a signal indicating the mass of material captured |
| | Inductive systems using measuring coils to assess the amount of ferrous material in circulation |
| | Measurement of pressure drop across the main full flow filter |
| **ON LINE** | |
| Monitoring a by-passed portion of the main oil flow | Optical measurement of turbidity as an indicator of particle concentration |
| | Pressure drop across filters of various pore sizes to indicate particle size distribution |
| | Discoloration of a filter strip after the passage of a fixed sample volume |
| | Resistance change between the grid wires of a filter to indicate the presence of metallic particles |
| **OFF LINE** | |
| Extracting a representative sample from the oil volume and analysing it remotely from the machine | Spectrometric analysis of the elemental content of the wear debris in order to determine its source |
| | Magnetic gradient separation of wear particles from a sample to determine their relative size, as a measure of problem severity |
| | Microscopic examination of the shape and size of the particles to determine the wear mechanisms involved |
| | Inductive sensor to give a direct numerical measurement of the level of ferrous debris in a sample of oil |
| | Optical particle counting on a diluted oil sample |

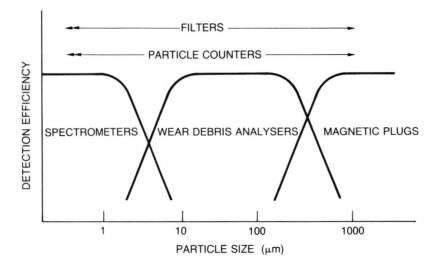

**Figure 14.1 The relative efficiency of various wear debris monitoring methods**

## Table 14.2 Off-line wear debris analysis techniques

| Technique | Description | Application |
|---|---|---|
| Atomic absorption spectroscopy | Oil sample is burnt in a flame and light beams of a wave length characteristic of each element are passed through the flame. The amount of light absorbed is a measure of the amount of the element that is present in the oil sample | Detects most common engineering metals. Detects particles smaller than $10\mu m$ only. Accurate at low concentrations of less than 5 ppm. |
| Atomic emission spectroscopy | Oil sample is burnt in an electric arc and the spectral colours in the arc are analysed for intensity by a bank of photomultipliers. Gives a direct reading of the content of many elements in the oil. | Detects most common engineering metals. Detects wear particles smaller than $10\mu m$ only. Accuracy is poor below 5 ppm. |
| Ferrography | A diluted oil sample is flowed across a glass slide above a powerful magnet. The particles deposit out on the slide with a distribution related to their size. | The size distribution can indicate the severity of the wear.<br><br>The particle shapes indicate the wear mechanism. |
| Rotary particle depositor | A diluted oil sample is placed on a glass cover above rotating magnets. Wear particles are deposited as a set of concentric rings. | As for ferrography with the added advantage that it can be linked directly to a particle quantifier |
| X-ray fluorescence | When the sample is exposed to a radioactive beam, X-rays characteristic of the material content are emitted. | Detects most engineering metals. Accurate to only $\pm 5$ ppm. |
| Inductively coupled plasma emission spectroscopy | The oil sample is sprayed into an argon plasma torch. The spectral colours of the emitted light and its intensity are then measured to indicate the amount of various elements that are present. | Detects most engineering metals.<br><br>Accurate down to parts per billion. |

## Table 14.3 Problems with wear debris analysis

| Problem | Solution |
|---|---|
| Poor quality or variable samples | Ensure that sample bottles are clean and properly labelled. Use sampling valves or suction syringes.<br><br>Hydraulic fluid sampling methods defined in ISO 3722. |
| Unrecorded oil change or a large oil addition | Often indicated by a sudden drop in contaminant levels.<br><br>The importance of recording oil top-ups needs to be emphasised to the operators. |
| Addition of the wrong oil to the machine detected by an increase of elements commonly used in oil additives | Take a second sample and discuss the problem with the machine operator in order to avoid a recurrence |

## Table 14.4 Sources of materials found in wear debris analysis

| Material | Likely worn component or other source | Material | Likely worn component or other sources |
|---|---|---|---|
| Aluminium | Light alloy pistons<br>Aluminium tin crankshaft bearings<br>Components rubbing on aluminium casings | Lead<br><br>Magnesium | Plain bearings<br><br>Wear of plastic components with talc fillers<br>Seawater intrusion |
| Antimony | White metal plain bearings | Nickel | Valve seats<br>Alloy steels |
| Boron | Coolant leaks<br>Can be present as an oil additive | Potassium | Coolant leaks |
| Chromium | Piston rings or cylinder liners<br>Valve seats | Silicon<br><br>Silver | Mineral dust intrusion<br><br>Silver-plated bearing surfaces<br>Fretting of silver soldered joints |
| Cobalt | Valve seats<br>Hard coatings | | |
| Copper | Copper-lead or bronze bearings<br>Rolling element bearing cages | Sodium<br><br>Tin | Coolant leakage<br>Seawater intrusion<br><br>Plain bearings |
| Indium | Crankshaft bearings | Vanadium | Intrusion of heavy fuel oil |
| Iron | Gears<br>Shafts<br>Cast iron cylinder bores | Zinc | A common oil additive |

## Table 14.5 Quick tests for metallic debris from filters

| Metal | Test method | Method | Test method |
|---|---|---|---|
| Steel and nickel | Can be attracted by a permanent magnet | Silver | Dissolves to form a white fog in nitric acid |
| Tin | Fuses with tin solder on a soldering iron | Copper and Bronze | Dissolves to produce a blue/green cloud in nitric acid |
| Aluminium | Dissolves rapidly in sodium or potassium hydroxide solution to form a white cloud | Chromium | Dissolves to produce a green cloud in hydrochloric acid |

## Physical characteristrics of wear debris

### Rubbing wear

The normal particles of benign wear of sliding surfaces. Rubbing wear particles are platelets from the shear mixed layer which exhibits super-ductility. Opposing surfaces are roughly of the same hardness. Generally the maximum size of normal rubbing wear is $15\mu$m.

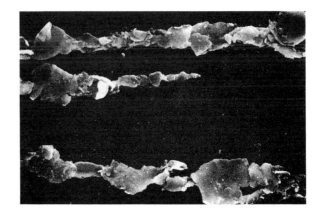

Break-in wear particles are typical of components having a ground or machined surface finish. During the break-in period the ridges on the wear surface are flattened and elongated platelets become detached from the surface often $50\mu$m long.

### Cutting wear

Wear particles which have been generated as a result of one surface penetrating another. The effect is to generate particles much as a lathe tool creates machining swarf. Abrasive particles which have become embedded in a soft surface, penetrate the opposing surface generating cutting wear particles. Alternatively a hard sharp edge or a hard component may penentrate the softer surface. Particles may range in size from $2-5\mu$m wide and 25 to $100\mu$m long.

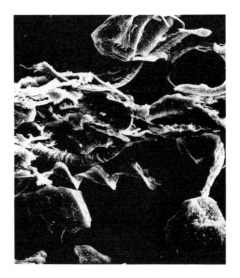

## Rolling fatigue wear

Fatigue spall particles are released from the stressed surface as a pit is formed. Particles have a maximum size of $100\mu m$ during the initial microspalling process. These flat platelets have a major dimension to thickness ratio greater than 10:1.

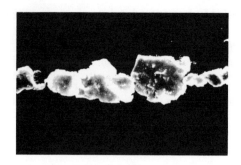

Spherical particles associated with rolling bearing fatigue are generated in the bearing fatigue cracks. The spheres are usually less than $3\mu m$ in diameter.

Laminar particles are very thin free metal particles between $20-50\mu m$ major dimension with a thickness ratio approximately 30:1. Laminar particles may be formed by their passage through the rolling contact region.

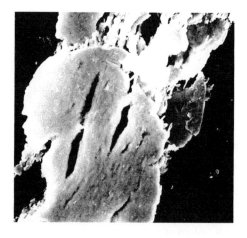

## Combined rolling and sliding (gear systems)

There is a large variation in both sliding and rolling velocities at the wear contacts; there are corresponding variations in the characteristics of the particles generated. Fatigue particles from the gear pitch line have similar characteristics to rolling bearing fatigue particles. The particles may have a major dimension to thickness ratio between 4:1 and 10:1. The chunkier particles result from tensile stresses on the gear surface causing fatigue cracks to propagate deeper into the gear tooth prior to pitting. A high ratio of large ($20\mu m$) particles to small ($2\mu m$) particles is usually evident.

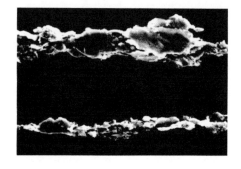

## Severe sliding wear

Severe sliding wear particles range in size from $20\mu m$ and larger. Some of these particles have surface striations as a result of sliding. They frequently have straight edges and their major dimension to thickness ratio is approximately 10:1.

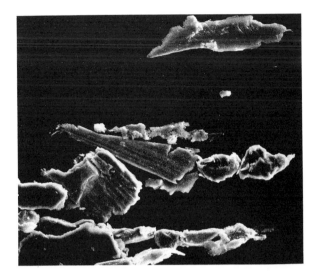

## Crystalline material

Crystals appear bright and changing the direction of polarisation or rotating the stage causes the light intensity to vary. Sand appears optically active under polarised light.

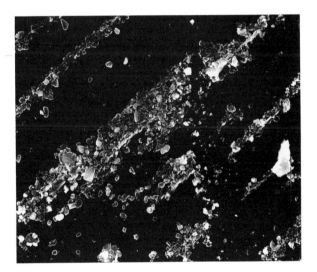

## Weak magnetic materials

The size and position of the particles after magnetic separation on a slide indicates their magnetic susceptibility. Ferro-magnetic particles (Fe, Co, Ni) larger than $15\mu m$ are always deposited at the entry or inner ring zone of the slide. Particles of low susceptibility such as aluminium, bronze, lead, etc, show little tendency to form strings and are deposited over the whole of the slide.

## Polymers

Extruded plastics such as nylon fibres appear very bright when viewed under polarised light.

## Examples of problems detected by wear debris analysis

### *Crankshaft bearings from a diesel engine*

Rapid wear of the bearings occurred in a heavy duty cycle transport operation. The copper, lead and tin levels relate to a combination of wear of the bearing material and its overlay plating.

| Sample no. | 1 | 2 | 3 | 4 |
|---|---|---|---|---|
| Iron ppm | 35 | 40 | 42 | 40 |
| Copper ppm | 15 | 25 | 35 | 45 |
| Lead ppm | 20 | 28 | 32 | 46 |
| Tin ppm | 4 | 4 | 10 | 15 |

### *Grease lubricated screwdown bearing*

The ratio of chromium to nickel, corresponding broadly to that in the material composition, indicated severe damage to the large conical thrust bearing.

| Sample no. | 1 | 2 | 3 | 4 |
|---|---|---|---|---|
| Iron ppm | 150 | 240 | 1280 | 1540 |
| Chromium ppm | 1 | 2 | 11 | 31 |
| Nickel ppm | 2 | 4 | 23 | 67 |

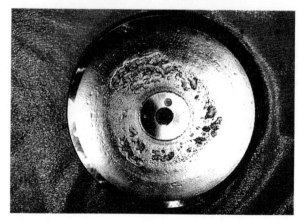

## Differential damage in an Intercity bus

Excessive iron and the combination of chromium and nickel resulted from the disintegration of a nose cone bearing

| Sample no. | 1 | 2 | 3 | 4 |
|---|---|---|---|---|
| Iron ppm | 273 | 383 | 249 | 71000 |
| Chromium ppm | 2 | 3 | 2 | 21 |
| Nickel ppm | 0 | 1 | 0 | 5 |

## Large journal bearing in a gas turbine pumping installation

The lead based white metal wore continuously.

| Sample no. | 1 | 2 | 3 | 4 |
|---|---|---|---|---|
| Iron ppm | 0 | 2 | 2 | 2 |
| Copper ppm | 4 | 5 | 11 | 15 |
| Lead ppm | 10 | 24 | 59 | 82 |

## Piston rings from an excavator diesel engine

Bore polishing resulted in rapid wear of the piston rings. The operating lands of the oil control rings were worn away.

| Sample no. | 1 | 2 | 3 | 4 |
|---|---|---|---|---|
| Iron ppm | 5 | 5 | 9 | 10 |
| Chromium ppm | 15 | 25 | 40 | 120 |

## Engine cylinder head cracked

The presence of sodium originates from the use of a corrosion inhibitor in the cooling water. A crack was detected in the cylinder head allowing coolant to enter the lubricant system.

| Sample no. | 1 | 2 | 3 | 4 |
|---|---|---|---|---|
| Iron ppm | 60 | 84 | 104 | 203 |
| Chromium ppm | 7 | 12 | 12 | 53 |
| Nickel ppm | 7 | 10 | 12 | 27 |
| Sodium ppm | 19 | 160 | 330 | 209 |

## LUBRICANT ANALYSIS

### Table 14.6 Off-line lubricant analysis techniques

| Technique | Description | Interpretation |
|---|---|---|
| Viscosity measurement | Higher viscosity than a new sample<br><br>Lower viscosity than a new sample | Oxidation of the oil and/or heavy particulate contamination<br><br>Fuel dilution in the case of engine oils |
| Total acid number, TAN<br>ASTM  D974<br>          D664<br>IP       139<br>          177 | Indicates the level of organic acidity in the oil | A measure of oil oxidation level<br><br>For hydraulic oils the value should not exceed twice the level for the new oil |
| Strong acid number, SAN<br>ASTM  D974<br>          D664<br>IP       139<br>          177 | Indicates the level of strong inorganic acidity in the oil | In engine oils indicates the presence of sulphur-based acids from fuel combustion. Synthetic hydraulic oils can show high values if they deteriorate |
| Total base number, TBN<br>ASTM  D664<br>          D2896<br>IP       177<br>          276 | Indicates the reserves of alkalinity present in the oil | Running engines on higher sulphur fuels creates acids which are neutralised by the oil, as long as it continues to be alkaline |
| Infra-red spectroanalysis | Measures molecular compounds in the oil such as water, glycol, refrigerants, blow by gases, liquid fuels, etc. Also additive content. | A very versatile monitoring method for lubricant condition |

### Table 14.7 Analysis techniques for the oil from various types of machine

| ** essential<br>* useful | Diesel engine | Gasoline engine | Gears | Hydraulic systems | Air compressor | Refrigeration compressor | Gas turbine | Steam turbine | Trans-formers | Heat transfer |
|---|---|---|---|---|---|---|---|---|---|---|
| Spectro-chemical analysis | ** | ** | ** | ** | ** | ** | ** | ** | | |
| Infra-red analysis | ** | ** | * | * | * | ** | * | * | * | * |
| Wear debris quantifier | ** | ** | ** | * | * | * | * | * | | |
| Viscosity at 40°C | ** | ** | ** | ** | ** | ** | ** | ** | | * |
| Viscosity at 100°C | * | * | | | | | | | | |
| Total base number | ** | ** | | | | | | | | |
| Total acid number | | | * | ** | * | * | ** | * | * | ** |
| Water % | ** | ** | * | ** | * | | | ** | ** | |
| Total solids | * | * | | | | | | | * | * |
| Fuel dilution | ** | ** | | | | | | | | |
| Particle counting | | | | ** | | | * | | | |

A useful condition monitoring technique is to check the performance of components, and of complete machines and plant, to check that they are performing their intended function correctly.

## COMPONENT PERFORMANCE

The technique for selecting a method of monitoring is to decide what function a component is required to perform and then to consider the various ways in which that function can be measured.

### Table 15.1 Methods of monitoring the performance of fixed components for fault detection

| Component | Function | Monitoring method |
|---|---|---|
| Casings and frameworks | Rigid support and transfer of loads to foundations | Crack detection by:<br>Visual inspection<br>Dye penetrants<br>Ultrasonic tests<br>Eddy current probes<br>Magnetic flux<br>Radiography<br><br>Tests for deflection under a known applied load.<br>Visual checks for material loss by corrosion |
| Cold pressure vessels | The containment of fluid under pressure | Detection of external surface cracks by:<br>Visual inspection<br>Dye penetrants<br>Eddy current probes<br>Magnetic flux<br><br>Detection of internal cracks by:<br>Ultrasonic tests<br>Radiography<br>Boroscopes (when out of service)<br><br>Detection of loss of wall thickness by corrosion:<br>Ultrasonic tests<br>Electrochemical probes<br>Sacrificial coupons<br>Small sentinel holes (where permissible)<br><br>Detection of strain growth by acoustic emission |
| Boilers and thermal reactors | The heating of fluids and containment of pressure | Detection of surface cracks and leakage by visual inspection<br>Detection of hidden cracks by ultrasonic tests<br>Hammer testing of the shell<br><br>Chemical check of feed water and boiler water samples, to indicate likely corrosion or deposit build up |
| Nozzle blades | The profiled flow of fluids and transmission of forces | Checking for profile changes and integrity by boroscopes (when out of service) |
| Ducts | Guiding the flow of air or other gases | Detection of leaks by gas sniffer detectors<br>If gas is hot and carrying fine solids, partial blockages can be detected by:<br>Infrared thermography<br>Thermographic paints |
| Pipes | Guiding the flow of fluids under pressure | Checking for reduction in wall thickness by:<br>Ultrasonics<br>Corrosion coupons<br>Electrochemical probes<br>Sentinel holes |

**Table 15.2 Methods of monitoring the performance of moving components for fault detection**

| Component | Function | Monitoring method |
|---|---|---|
| Journal bearings | Locating rotating shafts radially with the minimum friction | Checking low friction operation by temperature measurement |
| Thrust bearings | Locating rotating shafts axially with the minimum friction | Checking low friction operation by temperature measurement. Checking location by shaft position probes |
| Shafts and couplings | Smooth rotation while transmitting torques | Stroboscopes for visual inspection while rotating |
| Seals | Allowing rotating shafts to enter pressurised fluid containments with minimum leakage | Checking for leakage: Visually via deposits Gas sniffer detectors Liquid leakage pools |
| Pistons and cylinders | The interchange of fluid pressure and axial force with minimum leakage | Listening for leakage at ultrasonic frequencies |
| I.C. engine combustion | The provision of regular power pulses to drive the crankshaft | Toothed wheel fitted to the free end of the crankshaft to check for even rotation and pulses |
| Belt drives | The smooth transmission of power by differential tension | Stroboscopes for visual inspection while rotating. Proximity probes to detect excess slackness |
| Springs | Allowing controlled deflection with increasing loads | Deflection measurement at known loads. Crack detection by visual inspection and dye penetrants |

**Table 15.3 Methods of monitoring the performance of machines and systems for fault detection**

| Machine/System | Function | Monitoring method |
|---|---|---|
| Hydraulic power systems | Converting fluid pressure and flow into mechanical power | Measuring the relationship between pressure and flow in the system |
| Pumps | Converting mechanical power into fluid pressure and flow | Measuring the relationship between delivery pressure and flow, to detect any deterioration in the internal components |
| Textile machines | The production of fabric from threads | Checking the fabric for any patterns which indicate inconsistencies in machine operation |
| Coal or ore crushing mills | Reducing the particle size of materials | Checking changes in the size distribution of the output material |
| Motor vehicle | Consuming fuel to provide transportation | Measuring the distance covered per unit of fuel consumption |
| Heat exchangers | Producing temperature changes between two fluid flow streams | Monitoring the relationship between flow rate and temperature change |
| Thermodynamic and chemical process systems | Using pressure, temperature and volume/flow changes to interchange energy or change materials | The measurement and comparison against datums of pressure, temperature and flow relationships in the system |
| Systems with automatic control | The control of system variables to obtain a required output | The monitoring of the control actions taken by the system in normal operation to determine extreme values or combinations which indicate a system fault.<br><br>Giving the system a control exercise designed to detect likely faults |

## BALL AND ROLLER BEARINGS

If there is evidence of pitting on the balls, rollers or races, suspect fatigue, corrosion or the passage of electrical current. Investigate the cause and renew the bearing.

If there is observable wear or scuffing on the balls, rollers or races, or on the cage or other rubbing surfaces, suspect inadequate lubrication, an unacceptable load or misalignment. Investigate the cause and renew the bearing.

## ALL OTHER COMPONENTS

Wear weakens components and causes loss of efficiency. Wear in a bearing may also cause unexpected loads to be thrown on other members such as seals or other bearings due to misalignment. No general rules are possible because conditions vary so widely. If in doubt about strength or efficiency, consult the manufacturer. If in doubt about misalignment or loss of accuracy, experience of the particular application is the only sure guide.

Bearings as such are considered in more detail below.

## JOURNAL BEARINGS, THRUST BEARINGS, CAMS, SLIDERS, etc.

### Debris

If wear debris is likely to remain in the clearance spaces and cause jamming, the volume of material worn away in intervals between cleaning should be limited to $\frac{1}{5}$ of the available volume in the clearance spaces.

### Surface treatments

Wear must not completely remove hardened or other wear resistant layers.

Note that some bearing materials work by allowing lubricant to bleed from the bulk to the surface. No wear is normally detectable up to the moment of failure. In these cases follow the manufacturer's maintenance recommendations strictly.

Some typical figures for other treatments are:

| Treatment | Allowable wear depth |
|---|---|
| Good quality carburising | 2.5 mm (0.1 in) |
| Gas nitriding | 0.25 mm (0.01 in) |
| Salt bath nitriding | None |
| Cyanide hardening (shallow) | 0.025 mm (0.001 in) |
| Cyanide hardening (deep) | 0.25 mm (0.01 in) |
| Graphite or $MoS_2$ films | None |
| White metal, etc. | up to 50% of original thickness |

## Surface condition

Roughening (apart from light scoring in the direction of motion) usually indicates inadequate lubrication, overloading or poor surfaces. Investigate the cause and renew the bearing.

Pitting usually indicates fatigue, corrosion, cavitation or the passage of electrical current. Investigate the cause. If a straight line can be drawn (by eye) across the bearing area such that 10% or more of the metal is missing due to pits, then renew the components.

Scoring usually indicates abrasives either in the lubricant or in the general surroundings.

## Journal bearings with smoothly-worn surfaces

The allowable increase in clearance depends very much on the application, type of loading, machine flexibilities, importance of noise, etc., but as a general guide, an increase of clearance which more than doubles the original value may be taken as a limit.

Wear is gnerally acceptable up to these limits, subject to the preceding paragraphs and provided that more than 50% of the original thickness of the bearing material remains at all points.

## Thrust bearings, cams, sliders, etc. with smoothly-worn surfaces

Wear is generally acceptable, subject to the preceding paragraphs, provided that no surface features (for example jacking orifices, oil grooves or load generating profiles) are significantly altered in size, and provided that more than 50% of the original thickness of the bearing material remains at all points.

## CHAINS AND SPROCKETS

For effects of wear on efficiency consult the manufacturers. Some components may be case-hardened in which case data on surface treatments will apply.

## CABLES AND WIRES

For effects of wear on efficiency consult the manufacturer. Unless there is previous experience to the contrary any visible wear on cables, wires or pulleys should be investigated further.

## METAL WORKING AND CUTTING TOOLS

Life is normally set by loss of form which leads to unacceptable accuracy or efficiency and poor surface finish on the workpiece.

The surfaces of most components which have been worn, corroded or mis-machined can be built up by depositing new material on the surface. The new material may be applied by many different processes which include weld deposition, thermal spraying and electroplating.

The choice of a suitable process depends on the base material and the final surface properties required. It will be influenced by the size and shape of the component, the degree of surface preparation and final finishing required, and by the availability of the appropriate equipment, materials and skills. The following tables give guidance on the selection of suitable methods of repair.

### Table 17.1 Common requirements for all processes

| Requirement | Characteristic |
| --- | --- |
| Re-surfacing should be cheaper than replacement with a new part | Larger components often tend to be the most suitable and usually can be repaired on site. Also, the design may be unique or changes may have rendered the part obsolete, making replacement impossible. If the worn area was not surfaced originally the replacement surface may give better performance than that obtained initially. Repair, and re-surfacing are conservationally desirable processes, saving raw materials and energy and often proving environmentally friendly. |
| Appropriate surface condition prior to coating | Whatever process is considered applicable the surfaces for treatment should be in sound metallurgical condition, i.e. free from scale, corrosion products and mechanically damaged or work-hardened metal. The part may have the remains of a previous coating and unless it is possible to apply the new coating on top of this the old coating must be removed, generally by machining or grinding, sometimes by grit-blasting, anodic dissolution or heat treatment. Surfaces which have been nitrided, carburised, etc., may possibly be modified by heat treatment, otherwise mechanical removal will be necessary. |
| Preparatory machining of the surface prior to coating | If the component has to be restored to its original dimensions it will need to be undercut to allow for the coating thickness. In many cases this undercutting should be no more than $250\,\mu m/500\,\mu m$ deeper than the maximum wear which can be tolerated on the new surface. With badly damaged parts it will be necessary to undercut to the extent of the damage and this may limit the choice of repair processes which could be used. For precision components which require finishing by machining it is important to preserve reference surfaces to ensure that the finished component is dimensionally correct. |

# Repair of worn surfaces

## Table 17.2 Factors affecting the choice of process

| *Factor* | *Effect* |
|---|---|
| **SIZE AND SHAPE** Size and weight of component Thickness of deposit Weight to be applied Location of surfaces to be protected Accessibility | For small parts hand processes suitable. Spraying is a line-of-sight process limiting coating in bores, but some processes will operate in quite small diameters. Electroplating requires suitable anodes. Generally no limitation with weld deposition other than access of rod or torch in bores, however, some processes do not permit positional welding. |
| **DESIGN OF ASSEMBLY** Has brazing, welding, interference fits, riveting been used in the assembly? Are there temperature sensitive areas nearby? Have coaxial tolerances to be maintained? | Some processes put very little heat into an assembly so brazing filler metals would not be melted. Welds need careful cleaning and smoothing. Consideration must be given to the effect of heat on fitted parts. Suitable reference points are essential if tolerances are critical. |
| **DISTORTION PROBLEMS** What distortion is permissible? Is the part distorted as received? Will preparation cause distortion? Will surfacing cause distortion? Can distortion occur on cooling? | Eliminate distortion in part for repair. Remove cold-work stresses. Bond coats may replace grit-blasting. Some processes do not require surface roughening. Some processes provide more severe temperature gradients giving greater distortion. Allowance must be made for thermal expansion. Slow cooling must be available. In some cases pre-setting before coating may be used. Straightening or matching after coating may also be possible. |
| **PROPERTIES OF BASE METAL** Can chemical composition be determined? Have there been previous surface treatments/coating? What is surface condition? What is the surface hardness? What metallographic structure and base metal properties are ultimately required? | The depth of penetration and degree of surface melting can eliminate surface treatments, but previous removal may be necessary. If too hard for grit-blasting softening may be possible or a bond coat used. Suitable post-surfacing heat treatment may be needed. |
| **DEPOSIT REQUIREMENTS** Dimensional tolerances acceptable Surface roughness allowable/desirable Need for machining Further assembly to be carried out Additional surface treatments required, e.g. plating, sealing, painting, polishing Possible treatment of adjacent areas such as carburising Thermal treatments needed to restore base metal properties | Arc welding not suitable for thin coatings. To retain dimensions, a low or uniform heat input essential. If machining is needed a smooth deposit saves time and cost. Further assembly demands a dense deposit as do many surface treatments. Painting may benefit from some surface roughness, lubrication from porosity, some surfaces need to be smooth, low friction, others need frictional grip. These are some of the many factors needing consideration in selecting the process. |

**EFFECT ON BASE METAL PROPERTIES**

Consideration must be given to the effect of the process on base metal strength. Some sprayed coatings can reduce base metal fatigue strength as can hard, low ductility fused deposits in some environments. The same applies to some electro-deposited coats. Such coatings on high UTS steels can produce hydrogen embrittlement. Applied coatings do not usually contribute to base metal strength and allowance must be made for this, particularly where wear damage is machined out, prior to reclamation of the part. Welding will produce a heat affected zone which must be assessed. Fusing sprayed deposits can cause metallurgical changes in the base metal and the possibility of restoring previous properties must be considered. Unless coatings can be used as deposited, finish machining must be employed. Some coatings, e.g. flame sprayed or electro-deposited coatings must be finish ground, many dense adherent coatings may be machined. If porosity undesirable the coating must be sealed or densified.

**OTHER CONSIDERATIONS**

The number of components to be treated at one time or the frequency of repairs required, influence the expenditure permissible on surfacing equipment, handling devices, manipulators and other equipment.

The re-surfacing of equipment in situ such as quarry equipment, steelmill plant, paper making machinery, military equipment and civil engineering plant may limit or dictate choice of process.

## Table 17.3 The surfacing processes that are available

| Type | Symbol | Process | Coating materials and their form | Energy source and other materials | Surface preparation and coating procedure | Coating thickness | Treatment after coating |
|---|---|---|---|---|---|---|---|
| Gas Welding | GW | Gas welding | Rods, wires, tungsten carbide containing rods | Oxygen, acetylene | Flux application may be necessary if substrate contains elements forming refractory oxides, e.g. Cr | Up to 3 mm | Slow cooling / Machining / Grinding / Flux removal |
| Gas Welding | PW | Powder welding | Self-fluxing alloy powder | Oxygen, acetylene. Possibly flux | Flux application to high Cr steels. Pre-heating in some cases | Up to 5 mm | Slow cooling / Machining / Grinding / Filing softer deposits |
| Arc Welding AC and/or DC | MMA | Arc welding Manual (stick) welding | Bare rods, pre-alloyed wire electrodes or with alloying in the coating or tungsten carbide containing tubes | AC and/or DC Flux | May need pre-setting and/ or pre-heating. If wear excessive, preliminary build-up with low alloy steel economic | Thick coatings possible | Possibly slow cooling. If suitable heating treatment. If required straighten and/or machine. Generally flux/slag removal |
| Arc Welding AC and/or DC | TIG | Tungsten inert gas | Bare rods or wire | Argon. Tungsten electrodes | May need pre-setting and/ or pre-heating. May use manipulator | 2 mm upwards | Possibly slow cooling/heat treatment. Straightening, stress relieving, machining |
| Arc Welding AC and/or DC | MIG | Metal inert gas, gas shielded arc | Rods, wires or tubes | Carbon dioxide | As TIG | 3 mm upwards | As TIG |
| Arc Welding AC and/or DC | FCA | Flux-cored arc | Tubes, cored wires | | As TIG | 3 mm upwards | As TIG, flux removal may be necessary |
| Arc Welding AC and/or DC | SA | Submerged arc | Wires, strips | Granulated flux | As TIG. Will need use of an automatic manipulator | 3 mm upwards | As TIG. Flux removal needed |
| Arc Welding AC and/or DC | PTA | Plasma transferred arc | Metal or alloy powder | Argon, hydrogen and/or nitrogen, helium | As TIG. Use of manipulator/robot general | 2–5 mm | As TIG |
| Plating | EP | Electroplating | Ingots, plates, wires | Electric DC. Controlled chemical solutions. Cr, Ni possibly Cu anodes | Generally smooth turning or grinding. Protect areas where plating not desired. If steel has tensile strength exceeding $1000 \, N/mm^2$ or fatigue strength is important shot peen | Wide range | Heat treat to reduce coating stresses / Machine, grind |

**Table 17.3** (continued)

| Type | Symbol | Process | Coating materials and their form | Energy source and other materials | Surface preparation and coating procedure | Coating thickness | Treatment after coating |
|---|---|---|---|---|---|---|---|
| Thermal spraying/ Metallising | MP | Powder spraying | Metal or ceramic powder | Oxygen, acetylene sometimes hydrogen. Compressed air | Grit-blasting with or without grooving, rough threading or application of bond coat. Mask area not to be blasted or sprayed. Warm to prevent condensation from flame. Heat to reduce contraction stresses or improve adhesion | 2 mm upwards | May machine, impregnate, seal, use as sprayed or densify mechanically or by HIPping |
| | MW | Wire spraying | Metal wire, ceramic rods, powder/plastic wire | Oxygen, acetylene sometimes hydrogen. Compressed air | As MP | 75 μm upwards | As MP |
| | MA | Arc spraying | Solid or tubular wires | Electric power. Compressed air | Generally as MP. Al bronze an additional bond coat. Adjustment of spray conditions may provide coarse adherent deposit as initial bond coat. Mechanical handling often used. Warming/heating not used | 2 mm upwards | Use as sprayed, machine, seal |
| | HVOF | High velocity oxy/flame spraying | Metal powder, sometimes ceramic powder | Oxygen, composite gas, acetylene, hydrogen. Carrier gas Ar, He or N₂ | As MP | Wide range | As MP |
| | PF | Plasma flame spraying | Metal powder, perhaps wire or rod | Electric power. Argon, hydrogen nitrogen helium | As MP | Up to 2 mm | As MP |
| Sprayed and fused | SW | Sprayed and fused coatings | Self-fluxing alloy powders, occasionally bonded into wire | Oxygen, acetylene, (hydrogen), compressed air. Pre-heating torches/ burners | Use only chilled iron or nickel alloy grit which must be clean and free from fines. Do not use bond coats. If needed pre-set/ pre-heat. Mask areas not to be blasted/sprayed. Often lathe mounted and thermal expansion must be accommodated | 0.4–2.0 mm | Fuse. Slow cool. Possibly heat treat. Use as fused or machine |

**Table 17.4 Characteristics of surfacing processes**

| Process | Advantages | Disadvantages |
| --- | --- | --- |
| Gas welding | Small areas built up easily with thick deposits, grooves and recesses can be filled accurately. Thin, smooth coatings can be deposited. Minimum melting of the parent metal is possible with low dilution of the surfacing alloy. This is advantageous if using highly alloyed consumables or if a thin coating only desired. The process is under close control by the operator. Equipment is inexpensive and requires only fuel gases and possibly flux. A wide range of consumables available. There is minimum solution of carbon/carbide granules from tubular rods. Self-fluxing alloys can be deposited without flux, except on highly alloyed substrates, simplifying post-welding cleaning. | Process slow and not suitable for surfacing large areas. Build up of heat may overheat component and lead to distortion. Careful control of flame adjustment. As it is a manual process results are dependent on operator skill, fitness and degree of fatigue. Good technique is essential to ensure sound bond with the interface, especially if fusion with the substrate is not involved. There is a lack of NDT methods to check adhesion between coating and base metal. |
| Powder welding | Requires less skill than gas welding. Equipment not expensive but the self-fluxing powders are. The consumables have a wide temperature range between liquidus and solidus and between these temperatures the pasty consistency of the deposit enables thin, sharp edges and worn corners to be built up. Once an initial, thin coating has been built up high deposition rates of subsequent coatings are easily maintained. The ability to put down a thin first coating makes it possible to coat small areas on large components. Smooth or contoured deposits are possible requiring little finish machining. Soft deposits may be finished by hand filing. The process leaves one hand free to manipulate the work. | As for gas welding. Additionally there is a limited range of suitable consumables. It is essential that the interface temperature reaches 1000°C to ensure that the coating is bonded to and not just cast on the base metal and that there is enough heat and time for the self-fluxing action of the coating alloy to clean the surface of interfering oxides. |
| Manual metal arc welding | Equipment cost low, requiring very little maintenance. It is adaptable to small or large complex parts, can be used with limited access and positional welding possible, i.e. vertical. A wide range of consumables available. Deposition rates up to 5 kghr$^{-1}$. Ideal for one off and small series work and is useful when only small quantities of hardfacing alloys are required. | A skilled operator is needed for high quality deposits and slag removal is necessary. Dilution tends to be high. Granular carbides in tubular electrodes are usually melted. |
| TIG welding | The process can be closely controlled by the operator using hand-held torch and hand-held filler rod but can be mechanised for special applications. Small areas can be surfaced e.g. small pores in hard surfacing deposits. A deposit thickness of 2 mm upwards is achievable and a deposition rate up to 2 kghr$^{-1}$. High quality deposits can be made. | The process is slow and unsuitable for surfacing large areas. A limited range of consumables is available. The equipment is expensive and not suitable for site work, needing a workshop in which the shielding gas can be protected from disturbing draughts. |
| MIG welding | A continuous process which is used semi-automatically with a hand-held gun or is wholly mechanised by traversing the gun and/or the workpiece. Slag removal is not needed. It provides a positional surfacing capability and guns are available for internal bore work. High deposition rates 3–8 kghr$^{-1}$ with deposit thickness 3 mm upwards. | Equipment relatively expensive requiring regular maintenance. Use of shielding gas makes the process marginally less transportable than MMA and the gas must be selected to suit the surfacing alloy. Alloys for hard surfacing are not generally available in wire form so consumables restricted to mild steel (for build-up), stainless steels, aluminium bronze or tin bronze. High levels of UV radiation produced especially using high peak current pulse welding. |
| Flux-cored arc welding | Similar in principle to MIG surfacing but uses a tubular electrode containing a flux which decomposes to provide a shield to protect the molten pool. A separate shielding gas supply is not required. It is a continuous process used semi-automatically with a hand-held gun or is wholly mechanised. A wide range of consumables is available. High deposition rates up to 8 kghr$^{-1}$ for CO$_2$ shielded process and 11 kghr$^{-1}$ higher for self-shielded process. | With dilution of 15–30%, depending on technique, the process is not suitable for non-ferrous surfacing alloys. Equipment is fairly expensive and needs regular maintenance. Deposit quality may be lower than with MIG surfacing. Restrictions on use and transportation similar to the MIG process. |

**Table 17.4** (continued)

| Process | Advantages | Disadvantages |
|---|---|---|
| Submerged arc | It is a fully automatic process providing high deposition rates 10 kghr$^{-1}$ upwards on suitable workpieces. Deposit thickness 3 mm and over. A wide range of consumables available giving high quality deposits of excellent appearance requiring minimum finishing. Slag removal easy. | Intended primarily for workshop use with a fixed installation the equipment is very expensive and needs regular maintenance. Applications are limited. Generally, to large cylindrical or flat components; there is limited access to internal surfaces or larger bores. |
| PTA surfacing | A mechanised process giving close control of surface profile with minimum finishing required. Penetration and dilution low. Deposit thickness in range 2–5 mm. Deposition rate higher than with TIG process: 3.5 kghr$^{-1}$. Torches available for surfacing small bores down to about 25 mm diameter up to 400 mm deep. Very useful for surfacing internal valve seats. Balancing the power input between the arc plasma and the transferred arc enables deposits to be made on a wide range of components varying greatly in thickness and base metal composition. | Equipment expensive and not readily portable. Process costs high requiring a very skilled operator. |
| Electroplating | As operating temperatures never exceed 100°C work should not distort or suffer undesirable metallurgical changes. Coatings dense and adherent to substrate, molecular bonding may be as strong as 1000 N/mm$^2$. Plating conditions may be adjusted to modify hardness, internal stress and metallurgical characteristics of the deposits. No technical limits to thickness of deposits but most applications require thin coatings. Areas not requiring build-up can be masked. Brush plating can be used on localised areas, the equipment is portable and can be taken to the work. Deposition rates can exceed those of vat plating and may reach 200–400 mmhr$^{-1}$. | The thickness of deposit is proportional to current density and plating time, seldom exceeding 75 μmhr$^{-1}$. As current density over workpiece surface is seldom uniform coatings tend to be thicker at edges and corners and thinner in recesses and the centre of large flat areas.<br><br>Although application of coatings is not confined to line-of-sight the ability to plate round corners may be limited, however anode design and location may assist. The size of the vat limits dimensions of the work. Brush plating is labour intensive and requires considerable skill. The electrolytes are expensive but small volumes generally only required. |
| Powder spraying | Low cost equipment which can be used by semi-skilled operators. The most useful flame spray process for high alloy and self-fluxing surfacing materials continuously fed to the pistol. Coatings can be provided of materials which cannot be produced as rods or wires. Spray rates relatively high. Can provide coatings of uniform, controlled thickness, ideal for large cylindrical components but also possible on small and/or irregular parts. Low heat input to the base material minimises distortion and adverse metallurgical changes. Machining if needed can be minimal. | Most suitable for workshop use with adequate dust extraction available. Requires rough surface preparation. A line-of-sight process with limited access to bores and limitations on deposit thickness. Provides only a mechanical bond to the substrate. Deposit porous and may need sealing or densification. Needs compressed air as well as combustion gases. Not suitable for ceramic spraying. |
| Wire spraying | Generally as with powder spraying but much more suitable for site work. Wire reels can be at considerable distance from the spraying torch making it possible to work inside large constructions. Spray rates are fast and a wide range of surfacing materials can be sprayed. The spraying can be mechanised. Large areas easily sprayed given adequate design of work station. Can give useful porosity, thin coatings (75–150 μm) easily applied. For corrosion resistance sealed Zn, Al, or ZnAl coatings generally better than multi-layer paint systems. Ceramic spraying possible using rods. | As with powder spraying but does not require controlled mesh size of the consumables. |

**Table 17.4 (continued)**

| Process | Advantages | Disadvantages |
|---|---|---|
| Arc spraying | Faster than gas spraying, providing extremely good bond and denser coatings. Either solid or tubular wires can be used giving a wide range of coating alloys. Using two different wires, composite or 'pseudo-alloy' coatings are possible with compositions/structures unobtainable by other means. Can be used for thick build-ups on cylindrical parts. Surface preparation not as critical as for flame spraying. | Equipment more expensive than for flame spraying. Weight of gun and great spraying rate makes mechanical handling desirable. Limited to consumables available in wire form. Density and adhesion lower than the plasma spraying or HVOF process. |
| High velocity oxyflame spraying | Produces dense, high bond strength coatings and thick, low stress coatings independent of coating hardness. Tight spray pattern allows accurate placement of the deposit. Low heat input to the component. Torches available for spraying internal surfaces down to 100 mm diameter. Some equipments can spray ceramics. Adhesion is high eliminating need for a bond coat. | More expensive than flame spraying the process needs careful control and monitoring. Not really suitable for manual spraying, manipulators should be used. Consumables used are in powder form. |
| Plasma spraying | The high temperature enables almost all materials to be sprayed giving high density coatings strongly bonded to the substrate, with low heat input to it. This can cause problems with differential contraction between coating and base metal. High density, high adhesion, low oxide coatings with properties reproducable to close limits are possible by spraying in a vacuum chamber back-filled with argon. Heat input to the component is then much greater. | Capital cost higher than gas arc spraying. A spray booth is desirable. Vacuum spraying requires pumping equipment with a suitable manipulator. The size of the chamber limits work dimensions. Process time slow. High heat input to the workpiece may cause distortion or metallurgical changes in the substrate. |
| Spray, fused coatings | The coating is dense, non-porous and metallurgically bonded to the substrate. Spraying thickness and uniformity easily monitored and controlled. Although restricted to Ni- and Co-base self-fluxing alloys a large range of these available with a wide spread of hardness and other properties. Often used 'as fused' the smooth accurate surface facilitates machining with minimum wastage of the surface alloy. Tungsten carbide particles often incorporated in the coating to enhance wear resistance. Torches may be held manually or mechanically manipulated. Coatings, generally 0.4–2 mm thick can be put on components varying greatly in size and shape. Fusing can be carried out manually, in furnaces (generally vacuum), by induction heating or using lasers. | A lot of heat is introduced into the part. The interface must reach about 1000°C to ensure metallurgical bonding at the interface. This can cause distortion and metallurgical changes in the base material. Cooling from above 1000°C can provide problems due to the differential thermal contaction of coating and substrate metal. Coatings on transformable steels can provide cracking problems. |

## Table 17.5 General guidance on the choice of process

| Process | Cost of equipment | Ease of operation | Operator skill | Applicability of the process to: | | | | | | | | | | | Heat input on surfacing | | | Suitable surfacing alloys |
| | | | | Large areas | Small areas | Thick deposits | Thin deposits | Thick sections | Thin sections | Massive work | Small parts | Edge build-up | Site work | Machined components | Before | During | After | |
| --- | --- | --- | --- | --- | --- | --- | --- | --- | --- | --- | --- | --- | --- | --- | --- | --- | --- | --- |
| Gas welding | Low | Easy | Medium | 3 | 1 | 1 | 2 | 3 | 1 | 3 | 1 | 2 | 1 | 2 | Fair | High diffuse | May be desirable | Many. Excellent for WC containing rods. |
| Powder welding | Low | Easy | Low | 3 | 1 | 2 | 1 | 2 | 1 | 2 | 1 | 1 | 1 | 1 | Low | Fairly high | None | Restricted to self-fluxing alloys. |
| Manual metal arc welding | Fairly low | Fairly easy | Fairly high | 2 | 2 | 1 | 3 | 1 | 3 | 1 | 3 | 3 | 1 | 3 | Fair | Very high, steep gradients | May be desirable | Wide range available. |
| TIG welding | Medium high | Moderately easy | Fairly high | 3 | 1 | 2 | 2 | 1 | 2 | 2 | 2 | 2 | 2 | 2 | Low | Very high, fairly confined | May be desirable | Wide range available. |
| MIG welding | Medium | Moderately easy | Medium | 2 | 3 | 2 | 3 | 1 | 3 | 2 | 3 | 3 | 2 | 3 | Low | Very high, steep gradients | May be desirable | Restricted to alloys available as wires. |
| Flux-cored arc welding | Medium | Moderately easy | Medium | 1 | 3 | 2 | 3 | 1 | 3 | 1 | 4 | 3 | 2 | 3 | Low | Very high, steep gradients | May be desirable | Needs cored wires. |
| Submerged arc welding | Very high | Moderately hard | Fairly high | 1 | 4 | 1 | 4 | 1 | 4 | 1 | 4 | 4 | 4 | 4 | Low | Very high, diffuse | Some | Needs suitable coating materials or flux additions. |
| PTA surfacing | High | Hard | High | 4 | 1 | 3 | 2 | 3 | 1 | 3 | 2 | 3 | 4 | 1 | Low | High, moderate gradient | May be desirable | Many alloys possible. |
| Electroplating | High | Fairly easy | Medium | 1 | 1 | 2 | 1 | 1 | 1 | 2 | 1 | 3 | 4 | 1 | Low | Low | Low | Generally Cr, Ni and Cu for resurfacing. |
| Powder spraying | Fairly low | Easy | Low | 2 | 2 | 4 | 1 | 1 | 1 | 2 | 2 | 4 | 1 | 1 | Very low | Low | None | Limited, other than self-fluxing alloys. |
| Wire spraying | Fairly low | Easy | Low | 1 | 2 | 4 | 1 | 1 | 1 | 2 | 2 | 4 | 1 | 1 | Very low | Low | None | Considerable range including bond coats. Some ceramic rods. |
| Arc spraying | High | Easy | Fairly low | 1 | 3 | 2 | 2 | 1 | 1 | 1 | 2 | 4 | 1 | 1 | Very low | Fairly low | None | Limited range of wires. |
| HVOF spraying | Fairly high | Moderately easy | Medium | 2 | 2 | 2 | 1 | 1 | 1 | 2 | 2 | 3 | 2 | 1 | Very low | Fairly low | None | Considerable range of metallic powders. Some equipments spray plastics. |
| Plasma spraying | High | Fairly hard | High | 3 | 1 | 3 | 1 | 2 | 1 | 2 | 1 | 3 | 4 | 1 | Low | Fairly low | None | Many alloys and ceramics, usually powder |
| Sprayed and fused coatings | Fairly low | Fairly easy | Medium | 2 | 2 | 3 | 1 | 3 | 1 | 3 | 2 | 2 | 3 | 1 | Low | Fairly low | Fairly high, uniform | Needs self-fluxing alloys. |

1, Very applicable; 2, Not very applicable; 3, Not really suitable; 4, Unsuitable.

**Table 17.6 Available coating materials**

| Group | Sub-group | Alloy system | Important properties | Process suitability |  |  |  |  |  |  |  |  |  |  |  |  |  |  | Typical applications |
|---|---|---|---|---|---|---|---|---|---|---|---|---|---|---|---|---|---|---|---|
|  |  |  |  | GW | PW | MMA | TIG | MIG | FCA | SA | PTA | EP | MP | MW | MA | HVOF | PF | SW |  |
| STEEL (up to 1.7% C) | Pearlitic | Low carbon | Crack resistant, low cost, good base for hard-surfacing | 1 |  | X |  |  |  |  |  |  |  | 1 | 1 |  |  |  | Build up to restore dimensions. Track links, rollers, idlers |
|  | Martensitic | Low, medium or high carbon. Up to 9% alloying elements | Abrasion resistance increases with carbon content, resistance to impact decreases. Economical | 1 |  | X |  | X | X | 1 |  |  |  | 1 | X |  |  |  | Bulldozer blades, excavator teeth, bucket lips, impellers, conveyor screws, tractor sprockets, steel mill wobblers, etc |
|  | High speed | Complex alloy | Heat treatable to high hardness | X |  | 1 |  |  |  |  |  |  |  |  |  |  |  |  | Working and conveying equipment |
|  | Semi-austenitic | Manganese chromium | Tough crack resistant. Air and work hardenable | X |  | 1 |  |  |  | 1 |  |  |  |  |  |  |  |  | Mining equipment, especially softer rocks |
|  | Austenitic | Manganese | Work hardening | X |  |  | 1 | 1 | 1 |  |  |  |  |  |  |  |  |  | Rock crushing equipment |
|  |  | Alloyed manganese | Work hardening, less susceptible to thermal embrittlement. Useful build-up |  |  | X |  | 1 | 1 |  |  |  |  |  |  |  |  |  | Build up normal manganese steel prior to application of other hard-surfacing alloys |
|  |  | Chromium nickel | Stainless, tough, high temperature and corrosion resistant (low carbon) | 1 |  | 1 | X | X | X | X |  |  |  | 1 | X |  |  |  | Furnace parts, chemical plant |
| IRON (above 1.7% C) | High chromium | Martensitic | Show improved hot hardness and increased abrasion resistance | 1 |  | X | 1 | 1 | 1 | 1 |  |  |  |  |  |  |  |  | Steelworks equipment, scraper blades, bucket teeth |
|  |  | Multiple alloy | Hardenable. Can anneal for machining and re-harden. Good hot hardness | 1 |  | X | 1 | 1 | 1 | 1 |  |  |  |  |  |  |  |  | Mining equipment, dredger parts |
|  |  | Austenitic | Wide plastic range, can be hot shaped, brittle. Oxidation resistant | 1 |  | X |  | 1 | 1 | 1 |  |  |  |  |  |  |  |  | Low stress abrasion and metal-to-metal wear. Agricultural equipment |
|  | Martensitic alloy | Chromium tungsten Chromium molybdenum Nickel chromium | Very good abrasion resistance, very high compressive strength so can resist light impact. Can be heat-treated. Considerable variation in properties between gas and arcweld deposits | X |  | X | 1 | 1 | 1 | 1 | 1 |  |  |  |  |  |  |  | Cutting tools, shear blades, rolls for cold rolling |
|  | Austenitic alloy | Chromium molybdenum Nickel chromium | Lower compressive strength and abrasion resistance. Less susceptible to cracking. Will work harden | X |  | X | 1 | 1 | 1 | 1 | 1 |  |  |  |  |  |  |  | Mixing and steelworks equipment, agricultural implements |

**Table 17.6** (continued)

| Group | Sub-group | Alloy system | Important properties | GW | PW | MMA | TIG | MIG | FCA | SA | PTA | EP | MP | MW | MA | HVOF | PF | SW | Typical applications |
|---|---|---|---|---|---|---|---|---|---|---|---|---|---|---|---|---|---|---|---|
| | | | | | | | | | | | | | | | | | | | |
| CARBIDE | Iron base | Tungsten carbides in steel matrix | Resistant to severe abrasive wear. Care required in selection and application | X | X | X | 1 | 1 | 1 | 1 | | | | | | | | | Rock drill bits. Earth handling and digging equipment. Extruder screw augers |
| | Cobalt base | Tungsten carbides in cobalt alloy matrix | Matrix gives improved high temperature properties and corrosion resistance | X | X | 1 | 1 | 1 | 1 | | 1 | | | | | | | X | Oil refinery components, etc |
| | Nickel base | Tungsten carbides in nickel alloy matrix | Matrix gives improved corrosion resistance | | X | X | | | | | X | | | | | | | X | Screws, pump sleeves etc. in corrosive environments |
| | Copper base | Tungsten carbides in copper alloy matrix | Often larger carbide particles to give cutting and sizing properties | X | | | | | | | | | | | | | | | Oil field equipment |
| NICKEL BASE | Nickel | | High corrosion resistance | 1 | 1 | 1 | 1 | | | | | X | 1 | 1 | | | | | Chemical plant. Bond costs for ceramics |
| | Nickel Chromium Boron | With Fe, Si and C; W or Mo may be added | Self-fluxing alloys available in wide range of hardnesses. Abrasion, corrosion, oxidation resistant. Can be applied as thin, dense, impervious layers. Metallurgical bond to substrate | X | X | 1 | 1 | | | | X | | 1 | 1 | 1 | 1 | 1 | X | Glass mould equipment, engineering components, chemical and petrochemical industries |
| | Nickel Chromium | With possibly C, Mo or W to improve hardness and hot strength. Fe modifies thermal expansion improves creep resistance | Relatively soft and ductile. Good hot gas corrosion resistance. Very good corrosion resistance | X | X | 1 | 1 | | 1 | | X | | | | | | | | High temperature engineering applications. Chemical industry. I.C. engine valves |
| | Nickel Iron molydenum | 60% Ni, 20% Mo, 20% Fe or 65% Ni, 30% Mo, 5% Fe | Resistant to HCl, also sulphuric, formic and acetic acids | 1 | 1 | 1 | 1 | X | X | | 1 | | | | | | | | Chemical plant |
| | Nickel Copper | Monel | Corrosion resistant | 1 | 1 | 1 | 1 | | X | | | | | 1 | 1 | | | | Chemical plant |
| COBALT | Cobalt Chromium Tungsten | About 30% Cr, increasing W and C increases hardness | Superlative high temperature properties. Wear, oxidation and corrosion resistant | X | X | X | X | X | 1 | | X | | | | | | 1 | | Oilwell equipment, steelworks, chemical engineering plant, textile machinery |
| | Self-fluxing | With B, Si, Ni | Modified to provide self-fluxing properties. Good abrasion, corrosion resistance | 1 | X | 1 | 1 | | | | 1 | | X | | 1 | 1 | 1 | X | Chemical and petrochemical industries. Extrusion screws |

**Table 17.6** (continued)

| Group | Sub-group | Alloy system | Important properties | GW | PW | MMA | TIG | MIG | FCA | SA | PTA | EP | MP | MW | MA | HVOF | PF | SW | Typical applications |
|---|---|---|---|---|---|---|---|---|---|---|---|---|---|---|---|---|---|---|---|
| COPPER BASE | Copper | | Electrical conductivity | 1 | | 1 | | | | | | X | X | X | 1 | | | | Electrical equipment, paper-working machinery |
| | Bronze | Aluminium manganese, tobin, phosphor, commercial Al-bronze excellent bond coat arc metallising | Resistance to frictional wear and some chemical corrosion. | 1 | | 1 | | | | | | | X | X | X | | | | Bearing shells, shafts, slides, valves, propellers, etc |
| | Brass | | Bearing properties, decorative finishes | 1 | | | | | | | | | X | X | | | | | Water tight seals. Electric discharge machining electrodes |
| CHROMIUM | | | Wear, corrosion resistant | | | | | | | | | X | | | | | | | Engineering components |
| MOLYBDENUM | | | High adhesion to base metal | | | | | | | | | | | X | 1 | | | | Bond coats. Engineering parts reclamation |
| ALUMINIUM | | | Corrosion and heat resistance | | | | | | | | | | 1 | 1 | X | | | | Steel structures. Furnace parts |
| ZINC | | | Corrosion resistance | | | | | | | | | | X | X | 1 | | | | Steel structures, gas cylinders, tanks, etc. HF shielding |
| ZINC/ ALUMINIUM ALLOYS | | | Corrosion resistance | | | | | | | | | | | X | X | | | | Steel structures |
| LEAD BASE | Lead | | Resistance to chemical attack | 1 | | | | | | | | 1 | | X | 1 | | | | Resistance to sulphuric acid. Radiation shielding |
| | Solder | Often 60/40% Pb/Sn | Joining | 1 | | | | | | | | | | X | | | | | Tinning surfaces for subsequent joining |
| TIN BASE | Tin | | High corrosion resistance | 1 | | | | | | | | X | | X | 1 | | | | Electrical contacts. Food industry plant |
| | Babbitt | Sn, Sb, Cu | Bearing alloys | 1 | | | | | | | | | | X | 1 | | | | Bearing shells |
| OXIDES | | Principally of Al, Zn, Cr, Ti and mixtures | High temperature oxidation resistance. Wear resistance | | | | | | | | | | | X | | 1 | X | | Pump sleeves, aerospace parts |
| REFRACTORY METALS | | W, Ti, Ta, Cr | Good, high temperature properties. Often develop stable, protective oxide films | | | | | | | | | | | 1 | | 1 | X | | Electrical contacts. High density areas, corrosion protection |
| CARBIDES/ BORIDES | | Cr. W. B. possibly + Co or Ni | Very high wear resistance | 1 | | X | 1 | | | | | | | 1 | | X | X | | Thin cutting edges. Wear resistant areas |
| COMPLEX CERAMICS | | Silicides, titanates, zirconates, etc | Wear, oxidation, and erosion resistance | | | | | | | | | | | 1 | | X | X | | Thermal barriers, coating equipment handling molten metal and glass |

Process suitability

X frequently used, 1 sometimes used.

85

### Table 17.7 Factors affecting choice of coating material

| Factor | Effect |
|---|---|
| **PROPERTIES REQUIRED OF DEPOSIT**<br>Function of surface. Nature of adjacent surface or rubbing materials. What is the service temperature? What is the working environment (corrosive, oxidising, abrasive, etc.) What coefficient of friction required? | Is porosity desirable or to be avoided? Sprayed coatings are porous; fused or welded coatings are impervious. Porosity can be advantageous if lubrication required. What compressive stresses are involved? A porous deposit can be deformed and detached from the substrate. What degree of adhesion to the substrate is necessary? In many applications corrosion protection is satisfactory with mechanical adhesion – in some cases a metallurgical, bond will be required. How important is macro-hardness, metallographic structure – this often affects abrasion resistance. Is there danger of electro-chemical corrosion? If abrasion resistance is required, is it high or low stress? Many different alloys are available and needed to meet the many wear conditions which may be encountered. Are similar applications known? |
| **PROPERTIES NEEDED IN ALLOY**<br>Physical<br>Chemical<br>Metallurgical<br>Mechanical | Thermal expansion, or contraction, compared with the base metal, affects distortion. Chemical composition relates to corrosion, erosion and oxidisation resistance. The metallurgical structure is very important, influencing properties such as abrasion resistance and frictional properties. Mechanical properties can be critical. It is necessary to assess the relative importance of properties needed such as resistance to abrasion, corrosion, oxidation, erosion, seizure, impact and to what extent machinability, ductility, thermal conductivity or resistance, electrical conductivity or resistance may be required. |
| **PROCESS CONSIDERATIONS**<br>Surfacing processes available. Surface preparation methods available and feasible; auxiliary services which can be used. Location and size of work. Thickness of deposit required. | Some consumables can be produced only as wires – or powders – or cast rods. Choice could be limited by the equipment in use and operator experience. Lack of specific surface preparation methods could limit choice of process and this, in turn, prohibits use of certain materials. Similar limitations could arise from lack of necessary pre-heating or post-heat treatment plant. Can resurfacing be carried out without dismantling the assembly? Can work be carried out on site? Are only small areas on a large part to be repaired? |
| **ECONOMICS**<br>How much cost will the repair bear? | Often much more important than a direct cost comparison – cost of the repair compared to the purchase of a new part – are many other features, e.g. saving in subsequent maintenance labour and material costs, value of the lost production which is avoided, lower scrap rate during subsequent processing, time saved in associated production units, improved quality of component and product, increased production rates, reduction in consumption of raw materials. |

# Repair of worn surfaces

### Table 17.8 Methods of machining electroplated coatings

| Deposited metal | Machining with cutting tools | Grinding |
|---|---|---|
| Chromium | Chromium is too hard | Grinding or related procedures are the only suitable process. Soft or medium wheels should be used at the highest speed consistent with the limits of safety. Coolant must be continuous and copious. Light cuts only—preferably not exceeding 0.0075 mm should be taken. Heavy cuts can cause cracking or splintering of the deposit |
| Nickel | High-speed steel is, in general, the most satisfactory material for cutting nickel. Tipped tools are not recommended. The shape of the tools should be similar to those used for steel but with somewhat increased rake and clearance. Nickel easily work hardens therefore tools must be kept sharp and well supported to ensure that the cut is continuous | To avoid glazing, use an open textured wheel with a peripheral speed of 25–32 m/s. Coolant as for grinding steel |

### Table 17.9 Bearing materials compatible with electroplated coatings

| Material | Excellent | Good | Avoid |
|---|---|---|---|
| Chromium | Copper–lead, lead–bronze, white metal, Fine grain cast iron | Rubber or plastic, water-lubricated. Soft or medium hard steel with good lubrication and low speed. Brass, gun metal | Hard steel, phosphor bronze,* light alloys* |
| Nickel | | Bronze, brass, gun metal, white metal | Ferrous metals, phosphor bronze |

*May be satisfactory in some conditions.

When using deposited metals in sliding or rotating contact with other metals, adequate lubrication must be assured at all times.

# Repair of worn surfaces

**Table 17.10 Examples of successful repairs**

| Application | Coating material | Process used |
|---|---|---|
| Cast iron glass container moulds | Ni base self fluxing alloy in rod form<br>Ni base alloy | Gas welding<br>Powder welding |
| Steel plungers on molten glass pumps | Ni base alloy | Metal spraying |
| Excavator teeth | Bulk welding all over with chromium carbide alloy | Bulk welding |
| Excavator tooth tips | Hard alloy | Flux cored welding |
| Metering pump pistons | Aluminium oxide titanium oxide composite | Plasma flame spraying |
| Fire pump shafts | Stainless steel | Plasma wire spraying |
| Printing press rolls | Stainless steel on a nickel alumnide bond coat | Arc spraying |
| Large hydraulic press ram | Martensitic steel on a bond coat and sealed with a vinyl sealer | Wire spraying |
| Shafts of pumps handling slurries | Ni and Co base alloys with tungsten carbide | Plasma spraying |
| Screw conveyors | Ni base alloy rods | Gas welding |
| Small valves less than 12 cm diameter | Co–Cr–Ni–W alloys | Gas welding |
| Medium size valves | Co–Cr–Ni–W alloys | Powder welding |
| Very large valves | Co–Cr–Ni–W alloys | Submerged arc welding |
| Paper mill drums | Stainless steel wire | Arc spraying |
| Ammonia compressor pistons | Whitemetal | Wire spraying |
| Hydraulic pump plungers | Ni base alloy | Sprayed and fused |
| Guides on steel mills | Ni base alloy | Gas welding |

## ABRASIVE WEAR

Abrasive wear is the loss of material from a surface that results from the motion of a hard material across this surface.

There are several types of abrasive wear. Since the properties required of a wear-resistant material will depend on the type of wear the material has to withstand, a brief mention of these types of wear may be useful.

There are three main types of wear generally considered: gouging abrasion (impact), Figure 18.1; high-stress abrasion (crushing), Figure 18.2; and low-stress abrasion (sliding), Figure 18.3. This classification is made more on the basis of operating stresses than on the actual abrading action.

### Gouging abrasion

This is wear that occurs when coarse material tears off sizeable particles from wearing surfaces. This normally involves high imposed stresses and is most often encountered when handling large lumps.

### High-stress abrasion

This is encountered when two working surfaces rub together to crush granular abrasive materials. Gross loads may be low, while localised stresses are high. Moderate metal toughness is required; medium abrasion resistance is attainable.

Rubber now competes with metals as rod and ball mill linings with some success. Main advantages claimed are longer lifer at a given cost, with no reduction in throughput, lower noise level, reduced driving power consumption, less load on mill bearings and more uniform wear on rods.

### Low-stress abrasion

This occurs mainly where an abrasive material slides freely over a surface, such as in chutes, bunkers, hoppers, skip cars, or in erosive conditions. Toughness requirements are low, and the attainable abrasion resistance is high.

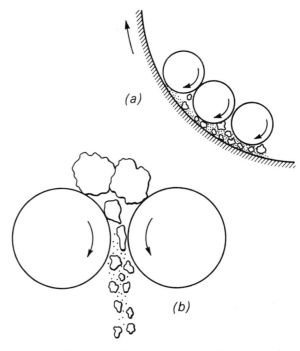

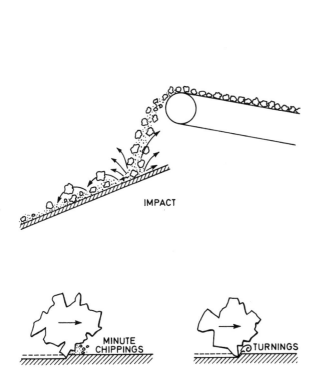

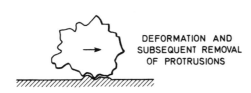

**Figure 18.2 Types of high-stress abrasion: (a) rod and ball mills; (b) roll crushing**

**Figure 18.1 Types of gouging abrasion**

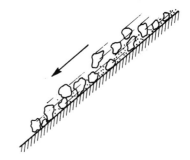

**Figure 18.3 Low-stress abrasion**

## MATERIAL SELECTION

Very generally speaking the property required of a wear-resistant material is the right combination of hardness and toughness. Since these are often conflicting requirements, the selection of the best material will always be a compromise. Apart from the two properties mentioned above, there are few *general* properties. Usually the right material for a given wear-resistant application can only be selected after taking into consideration other factors that determine the rate of wear. Of these the most important are:

Ambient temperature, or temperature of material in contact with the wear surface.
Size distribution of particles flowing over the wear surface.

Abrasiveness of these particles.
Type of wear to which wear surface is subjected (i.e. gouging, sliding, impact, etc.).
Velocity of flow of material in contact with wear surface.
Moisture content or level of corrosive conditions.
General conditions (e.g. design of equipment, head-room available, accessibility, acceptable periods of non-availability of equipment).

Tables 18.1 and 18.2 give some general guidance on material selection and methods of attaching replaceable components.

Table 18.3 gives examples of actual wear rates of various materials when handling abrasive materials.

The subsequent tables give more detailed information on the various wear resistant materials.

### Table 18.1 Suggested materials for various operating conditions

| Operating conditions | Properties required | Material |
|---|---|---|
| High stress, impact | Great toughness; work-hardening properties | Austenitic manganese steel, rubber of adequate thickness |
| Low stress, sliding | 1, Great hardness; 2, toughness less important; 3, quick replacement | Hardened and/or heat-treated metals, hardfacing, ceramics |
| | 1, Cheapness of basic material; 2, replacing time less important | Ceramics, quarry tiles, concretes |
| | 1, Maximum wear resistance; cost is immaterial | Tungsten carbide |
| Gouging wear | High toughness | Usually metals, i.e. irons and steels, hardfacing |
| Wet and corrosive conditions | Corrosion resistance | Stainless metals, ceramics, rubbers, plastics |
| Low stress; contact of fine particles; low abrasiveness | Low coefficient of friction | Polyurethane, PTFE, smooth metal surfaces |
| High temperature | Resistance to cracking, spalling, thermal shocks; general resistance to elevated temperatures | Chromium-containing alloys of iron and steel; some ceramics |
| Minimum periods of shut-down of plant | Ease of replacement | Any material that can be bolted in position and/or does not require curing |
| Curved, non-uniform irregular surface and shapes | Any one or a combination of the above properties | Hardfacing weld metal; most trowellable materials |
| Arduous and hot conditions | | Hardfacing weld metal |

## Table 18.2 Methods of attachment of replaceable wear-resistant components

| Method of fixing | | Suitable for: |
|---|---|---|

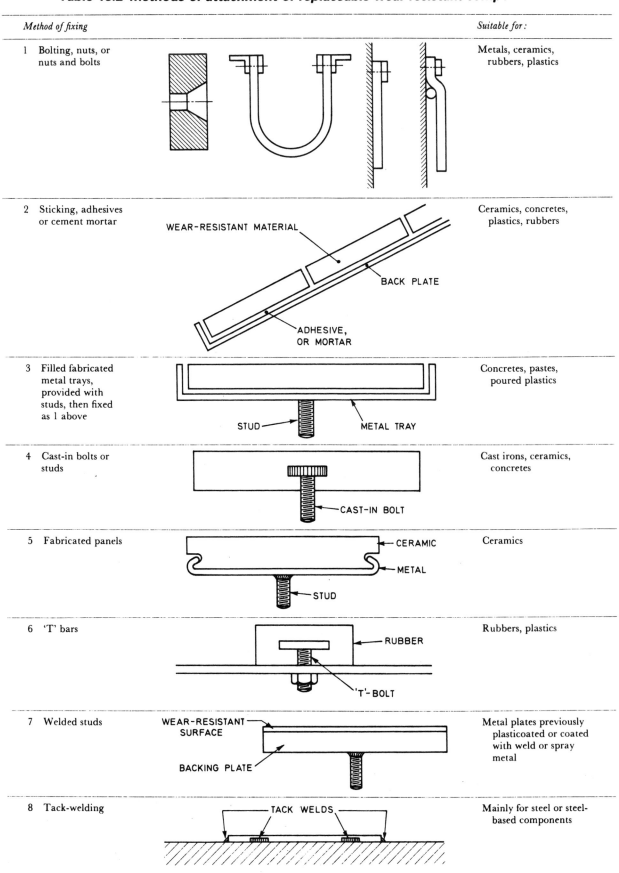

| | | |
|---|---|---|
| 1 Bolting, nuts, or nuts and bolts | | Metals, ceramics, rubbers, plastics |
| 2 Sticking, adhesives or cement mortar | WEAR-RESISTANT MATERIAL / BACK PLATE / ADHESIVE, OR MORTAR | Ceramics, concretes, plastics, rubbers |
| 3 Filled fabricated metal trays, provided with studs, then fixed as 1 above | STUD / METAL TRAY | Concretes, pastes, poured plastics |
| 4 Cast-in bolts or studs | CAST-IN BOLT | Cast irons, ceramics, concretes |
| 5 Fabricated panels | CERAMIC / METAL / STUD | Ceramics |
| 6 'T' bars | RUBBER / 'T'-BOLT | Rubbers, plastics |
| 7 Welded studs | WEAR-RESISTANT SURFACE / BACKING PLATE | Metal plates previously plasticoated or coated with weld or spray metal |
| 8 Tack-welding | TACK WELDS | Mainly for steel or steel-based components |

**Table 18.3 Typical performance of some wear-resistant materials as a guide to selection**

| Type | Some typical materials | Sliding wear-rate* by coke | Sliding wear-rate* by sinter | Temperature limitations | Ease and convenience of replacement | General comments |
|---|---|---|---|---|---|---|
| Cast irons | Ni-hard type martensitic white iron | 0.11 | 0.06 | | Yes | The most versatile of the materials which, now, by varying alloying elements, method of manufacture and application are able to give a wide range of properties. Their main advantage is the obtainable combination of strength, i.e. toughness and hardness, which accommodates a certain amount of abuse. Other products are sintered metal and metal coatings |
| | High chrome martensitic white irons | 0.12 | 0.11 | | | |
| | Spheroidal graphite-based cast iron | 0.22 | 0.09 | | | |
| | High phosphorus pig iron | 0.32 | 0.91 | | | |
| | Low alloy cast iron | – | 0.44 | | | |
| Cast steels | 3¼ Cr-Mo cast steel | 0.17 | – | | | |
| | 13 Mn austenitic cast steel | 0.22 | – | | | |
| | 1½ Cr-Mo cast steel | 0.43 | – | | | |
| Rolled steels | Armour plate | 0.12 | – | | | |
| | Work-hardened Mn steel | 0.13 | – | | | |
| | Low alloy steel plate, quenched and tempered | 0.31 | 0.30–0.84 | | | |
| | EN8 steel | 0.43 | 0.63 | | | |
| Hard facings | High chrome hardfacing welds, various | 0.09–0.16 | 0.05–0.14 | No | Could be difficult if applied *in situ* | Great range of hardness. Most suitable for low-stress abrasion by low-density materials, and powders. Disadvantage: brittleness |
| Ceramics | Fusion-cast alumina-zirconia-silica | 0.05 | 0.11–0.14 | | Yes, if bolted. Not so convenient if fixed by adhesive or cement mortar, as long curing times may be unacceptable | |
| | Slagceram | 0.15 | 0.33 | | | |
| | Fusion-cast basalt | 0.17 | 0.53 | | | |
| | Acid-resisting ceramic tile | 0.19 | 1.27 | | | |
| | Plate glass | 0.81 | – | | | |
| | Quarry floor tiles | 2.2–3.4 | – | | | |
| Concretes | Aluminous cement concrete. | 0.42 | 4.0–4.4 | | Could be messy. Might be difficult under dirty conditions | Advantages: cheapness, castability. Disadvantage: long curing or drying-out times |
| | Quartz–granite aggregate-based concrete | 0.87 | 6.5 | | | |
| Rubbers | Rubbers, various | 2.1–3.2 | – | | Bonded and bolted. Stuck with adhesive, could be difficult under dirty conditions | Main advantage is resilience and low density, with a corresponding loss in bulk hardness. The most useful materials where full advantage at the design stage can be taken of their resilience and anti-sticking properties |
| Rubber-like plastics | Polyurethanes, various | 2.3–5.4 | 2.3 | Yes | | |
| Other plastics | High-density polyethylene | 6.4 | – | | In sheet form it is difficult to stick | Low coefficient of friction, good antisticking properties. Best for low-stress abrasion by fine particles |
| | Polytetrafluoroethylene (PTFE) | 8.2 | – | | | |
| Resin-bonded compounds | Resin-bonded calcined bauxite | 2.3 | – | | Trowelled; could be messy. Difficult in dirty and inaccessible situations | These materials are only as strong as their bonding matrix and therefore find more application where low-stress wear by powders or small particles (grain, rice) takes place |

*Wear rate is expressed in in³ of material worn away per 1000 tons of the given bulk materials per ft² of area in contact with the abrading material. The results were obtained from field trials in a chute feeding a conveyor belt.

This data is provided as examples of the relative wear rates of the various materials when handling abrasive bulk materials.

The following tables give more detailed information on the materials listed in Table 18.3 with examples of some typical applications in which they have been used successfully.

When selecting the materials for other applications, it is important to identify the wear mechanism involved as this is a major factor in the choice of an optimum material. Further guidance on this is given in Table 18.1.

### Table 18.4 Cast irons

| Type | Nominal composition | Hardness Brinell | Characteristics | Typical application |
|------|--------------------|-----------------|-----------------|--------------------|
| Grey irons BS 1452 ASTM A48 | Various | 150–300 | Graphite gives lubrication | Brake blocks and drums, pumps |
| Spheroidal graphite | Meehanite WSH2 | Up to 650 | Heat-treatable. Can be lined with glass, rubber, enamels | Many engineering parts, crusher cones, gears, wear plates |
| High phosphorus | 3.5%C 2.0%P | Up to 650 | Brittle, can be reinforced with steel mesh. | Sliding wear |
| Low alloy cast iron | 3%C 2%Cr 1%Ni | 250–700 | | Sliding wear, grate bars, cement handling plant, heat-treatment |
| NiCr Martenstic irons Typical examples: NiHard, BF 954 | 2.8–3.5%C, 1.5–10% Cr 3–6% Ni | 470–650 | | High abrasion Ore handling, sand and gravel |
| Ni Hard 4 | 7–9% Cr, 5–6.5% Ni, 4–7% Mn | | | More toughness. Resists fracture and corrosion |
| CrMoNi Martensitic irons Typical examples: Paraboloy | 14–22% Cr, 1.5% Ni, 3.0% Mo | 500–850 | | Ball and rod mills, wear plates for fans, chutes, etc. |
| High Chromium irons Typical examples: BF 253 HC 250 | 22–28% Cr | 425–800 | Cast as austenite Heat-treated to martensite | Crushing and grinding Plant Ball and Rod mills Shot blast equipment, pumps |

# Wear resistant materials

### Table 18.5  Cast steels

| Type | Nominal composition | Hardness | Characteristics | Typical applications |
|---|---|---|---|---|
| Carbon steel BS 3100 Grade A | | Up to 250 | | Use as backing for coatings |
| Low alloy steels BS 3100 Grade B | Additions of Ni, Cr, Mo up to 5% | 370–550 | | For engineering 'lubricated' wear conditions |
| Austenitic BS 3100 BW10 | 11% Mn min. | 200 soft Up to 600 when work-hardened | | For heavy impact wear, Jaw and Cone crushers, Hammer mills |
| High alloy steels BS 3100 Grade C | 30% Cr 65% Ni + Mo, Nb etc. | 500 | | Special alloys for wear at high temperature and corrosive media |
| Tool steels Many individual specifications | 17% Cr, 4% Ni, 9% Mo, 22% W, 10% Co | Up to 1000 | | Very special applications, usually as brazed-on plates. |

### Table 18.6  Rolled steels

| Type | Nominal composition | Hardness | Characteristics | Typical application |
|---|---|---|---|---|
| Carbon steels BS 1449 Part 1 Typical examples: BS 1449 Grade 40 (En8 plate) Abrazo 60 | .06/1.0% C, 1.7% Mn | 160–260 | Higher carbon for low/medium wear | For use as backing for hard coatings |
| Low alloy steels Many commercial specifications Typical examples: ARQ Grades, Tenbor 25 30, Wp 300 and 500, Creusabro Grades, Abro 321 and 500, OXAR 320 and 450, Red Diamond 20 & 21, Compass B555 | Up to 3.5% Cr, 4% Ni, 1% Mo | 250–500 | Quenched and tempered. Are weldable with care | Use for hopper liners, chutes, etc. |
| Austenitic manganese steel Typical examples: Cyclops 11/14 Mn, Red Diamond 14 | 11/16% Mn | 200 in soft condition 600 skin hardened by rolling | | |
| High alloy and stainless steel BS 1449 Part 2 | Up to 10% Mn, 26% Cr, 22% Ni, + Nb, Ti | Up to 600 | Heat and corrosion resistant | Stainless steels |

## Table 18.7 Wear resistant coatings for steel

| | Method | Technique | Materials | Characteristics and applications |
|---|---|---|---|---|
| *Weld applied surfaces* | Gas welding | Manual | Rods of wide composition. Mainly alloys of Ni, Cr, Co, W etc. | For severe wear, on small areas. Thickness up to 3 mm |
| | Arc welding | Manual | Coated tubular electrodes. Specifications as above | Wear, corrosion and impact resistant. Up to 6 mm thickness. |
| | | Semi- or full automatic | Solid or flux-cored wire, or by bulk-weld Tapco process. Wide range of materials | As above, but use for high production heavy overlays 10–15 mm. |
| | Fused paste | Paste spread onto surface, then fused with oxy-fuel flame or carbon-arc. | Chromium boride in paste mix | Excellent wear resistance. Thin (1 mm) coat. Useful for thin fabrications: fans, chutes, pump impellors, screw conveyors, agricultural implements |
| *Flame spray processes* | Oxy-fuel | Consumable in form of powder, wire, cord or rod, fed through oxy-fuel gun. Deposit may be 'as-sprayed' or afterwards fused to give greater adhesion | Materials very varied, formulated for service duty | Wear, corrosion heat, galling, and impact resistance. Thicknesses vary depending on material. Best on cylindrical parts or plates |
| | Arc spray | Wire consumable fed through electric arc with air jet to propel molten metal | Only those which can be drawn into wire | High deposition rate, avoid dew-point problems, therefore suited to larger components |
| *Plasma spray processes* | Non-transferred | Similar to flame spray, but plasma generated by arc discharge in gun | Materials as for flame-spray, but refractory metals, ceramics and cermets in addition, due to very high temperatures developed | High density coatings. Very wide choice of materials. Application for high temperature resistance and chemical inertness |
| | Transferred | As above but part of plasma passes through the deposit causing fusion | Mainly metal alloys | High adhesion Low dilution Extremely good for valve seats |
| | Detonation gun | Patent process of Union Carbide Corp. Powder in special gun, propelled explosively at work | Mainly hard carbides and oxides | Very high density. Requires special facilities |
| | High velocity oxy-fuel | Development of flame gun, gives deposit of comparable quality to plasma spray | Similar to plasma spray | More economic than plasma |
| *Others* | Hard chromium plating | Electrode position | Hard chrome Up to 950 HV (70 Rc) | Wear, corrosion and sticking resistant |
| | Electroless nickel | Chemical immersion | Nickel phosphide 850 HV after heat treatment | Similar to hard chrome |
| | Putty or paint | Applied with spatula or brush | Epoxy or polyester resins, or self-curing plastics, filled with wear resistant materials | Wear and chemical resistant |

### Table 18.8 Some typical wear resistant hardfacing rods and electrodes

| Material type | Name | Typical application |
|---|---|---|
| Low alloy steels | Vodex 6013, Fortrex 7018, Saffire Range. Tenosoudo 50, Tenosoudo 75, Eutectic 2010 | Build-up, and alternate layering in laminated surfaces |
| Low alloy steels | Brinal Dymal range. Deloro Multipass range. EASB Chromtrode and Hardmat. Metode Met-Hard 250, 350, 450. Eutectic N6200, N6256. Murex Hardex 350, 450, 650, Bostrand S3Mo. Filarc 350, Filarc PZ6152/PZ6352. Suodometal Soudokay 242-0, 252-0, 258-0, Tenosoudo 105, Soudodur 400/600, Abrasoduril. Welding Alloys WAF50 range Welding Rods Hardrod 250, 350, 650 | Punches, dies, gear teeth, railway points |
| Martensitic chromium steels | Brinal Chromal 3, ESAB Wearod, Metrode Met-Hard 650. Murex Muraloy S13Cr. Filarc PZ6162. Oelikon Citochrom 11/13. Soudometal Soudokay 420, Welding Rods Serno 420FM. Welding Alloys WAF420 | Metal to metal wear at up to 600C. High C types for shear blades, hot work dies and punches, etc. |
| High speed steels | Brinal Dyma H. ESAB OK Harmet HS. Metrode Methard 750TS. Murex-Hardex 800, Oerlikon Fontargen 715. Soudometal Duroterm 8, 12, 20, Soudostel 1, 12, 21. Soudodur MR | Hot work dies, punches, shear blades, ingot tongs |
| Austenitic stainless steels | Murex Nicrex E316, Hardex MnP, Duroid 11, Bostrand 309. Metrode Met-Max 20.9.3, Met-Max 307, Met-Max 29.9 Soudometal Soudocrom D | Ductile buttering layer for High Mn steels on to carbon steel base. Furnace parts, chemical plant |
| Austenitic manganese steels | Brinal Mangal 2. Murex Hardex MnNi Metrode Workhard 13 Mn, Workhard 17 MnMo, Workhard 12MnCrMo. Soudometal Soudomanganese, Filarc PZ6358 | Hammer and cone crushers, railway points and crossings |
| Austenitic chromium manganese steels | Metrode Workhard 11Cr9Mn, Workhard 14Cr14Mn. Soudometal Comet MC, Comet 624S | As above but can be deposited on to carbon steels. More abrasion resistant than Mn steels |
| Austenitic irons | Soudometal Abrasodur 44. Deloro Stellite Delcrome 11 | Buttering layer on chrome irons, crushing equipment, pump casings and impellors |
| Martensitic irons | Murex Hardex 800. Soudometal Abrasodur 16. Eutectic Eutectdur N700 | For adhesive wear, forming tools, scrapers, cutting tools |
| High chromium austenitic irons | Murex Cobalarc 1A, Soudometal Abrasodur 35, 38. Oerlikon Hardfacing 100, Wear Resistance WRC. Deloro Stellite Delcrome 91 | Shovel teeth, screen plate, grizzly bars, bucket tips |
| High chromium martensitic irons | Metrode Met-hard 850, Deloro Stellite Delcrome 90 | Ball mill liners, scrapers, screens, impellors |
| High complex irons | Brinal Niobal. Metrode Met-hard 950, Met-Hard 1050. Soudometal Abrasodur 40, 43, 45, 46 | Hot wear applications, sinter breakers and screens |
| Nickel alloys | Metrode 14.75Nb, Soudonel BS, Incoloy 600. Metrode 14.75MnNb, Soudonel C, Incoloy 800. Metrode HAS C, Comet 95, 97, Hastelloy types | Valve seats, pump shafts, chemical plant |
| Cobalt alloys | EutecTrode 90, EutecRod 91 | Involving hot hardness requirement: Valve seats, hot shear blades |
| Copper alloys | Saffire Al Bronze 90/10, Citobronze, Soudobronze | Bearings, slideways, shafts, propellers |
| Tungsten carbide | Cobalarc 4, Diadur range | Extreme abrasion: fan impellors, scrapers |

## Table 18.9 Wear resistant non-metallic materials

| Type | Nominal composition | Hardness | Characteristics | Typical application |
|------|---------------------|----------|-----------------|---------------------|
| **Extruded ceramic** Indusco Vesuvius | High density Alumina | 9 Moh | Process limits size to $100 \times 300 \times 50$ mm. Low stress wear, also at high temperature | |
| **Ceramic plates** Hexagon-shaped, cast Indusco | | | | Suitable for lining curved surfaces |
| **Sintered Alumina** Alumina 1542 | $96\%Al_2O_3$ $2,4\%SiO_2$ | 9 Moh | Low stress wear also at high temperatures | |
| Isoden 90 | $90\%Al_2O_3$ | | | |
| Isoden 95 | $95\%Al_2O_3$ | | | |
| **Fusion-cast Alumina** Zac 1681 | $50\%Al_2O_3$ $32.5\%ZrO_2$ $16\%SiO_2$ | | Can be produced in thick blocks to any shape. Low stress and medium impact, also at high temperatures | |
| **Concrete** Alag Ciment Fondu | Mainly calcium silico-aluminates $40\%Al_2O_3$ | | Low cost wear-resistant material. High heat and chemical resistance | Floors, coke wharves, slurry conveyors, chutes |
| **Cast Basalt** Heat-treated | Remelted natural basalt | 7–8 Moh | Low stress abrasion. Brittle | Floors, coke chutes, bunker, pipe linings, usually 50 mm thick minimum. Therefore needs strong support |
| **Plate Glass** | | Glass | Very brittle | Best suited for fine powders, grain, rice etc. |
| **Rubber** Trellex Skega Linatex | Various grades 95% Natural rubber | Various Fairly soft | Resilient, flexible | Particularly suitable for round particles, water borne flow of materials |
| **Ceramic ballsheet** Hoverdale | Rubber filled with ceramic balls | | Enhanced wear resistance | |
| **Plastics** Duthane Flexane Tivarthane (Polyhi-Solidor) Scandurathane (Scandura) Supron (Slater) | Polyurethane based, rubber-like materials | | Low stress abrasion applications | Floors, chute liners screens for fine materials |
| Duplex PTFE | Polytetra-fluorethylene | | Low coefficient of friction | For fine powders light, small particles |
| **Resins** Belzona Devcon Greenbank AD1 Thortex Systems Nordbak | Resin-based materials with various wear-resistant aggregates | | Can be trowelled. Specially suitable for curved and awkward surfaces but not for lumpy materials | Floors, walls, chutes, vessels. In-situ repairs |

*Ceramic materials* (left margin)
*Rubber and plastics* (left margin)

In general, the repair of bearings by relining is confined to the low melting-point whitemetals, as the high pouring temperatures necessary with the copper or aluminium based alloys may cause damage or distortion of the bearing housing or insert liner. However, certain specialist bearing manufacturers claim that relining with high melting-point copper base alloys, such as lead bronze, is practicable, and these claims merit investigation in appropriate cases.

For the relining and repair of whitemetal-lined bearings three methods are available:
(1) Static or hand pouring.
(2) Centrifugal lining.
(3) Local repair by patching or spraying.

### Table 19.1 Guidance on choice of lining method

| Type of bearing | Relining method | Field of application |
|---|---|---|
| Direct lined housings | Static pouring or centrifugal lining | Massive housings. To achieve dynamic balance during rotation, parts of irregular shape are often 'paired' for the lining operation, e.g. two cap half marine type big-end bearings lined together, ditto the rod halves |
| INSERT LINERS | | |
| 'Solid inserts' | Not applicable | New machined castings or pressings required |
| Lined inserts | | |
| Thick walled | Static pouring or centrifugal lining | Method adopted depends on size and thickness of liner, and upon quantities required and facilities available |
| Medium walled | | |
| Thin walled | Not recommended | Relining not recommended owing to risk of distortion and loss of peripheral length of backing. If relining essential (e.g. shortage of supplies) special lining jigs and protective measures essential |

## (1) PREPARATION FOR RELINING

(a) Degrease surface with trichlorethylene or similar solvent degreaser. If size permits, degrease in solvent tank, otherwise swab contaminated surfaces thoroughly.
(b) Melt off old whitemetal with blowpipe, or by immersion in melting-off pot containing old whitemetal from previous bearings, if size permits.
(c) Burn out oil with blowpipe if surface heavily contaminated even after above treatment.
(d) File or grind any portions of bearing surface which remain contaminated or highly polished by movement of broken whitemetal.
(e) Protect parts which are not to be lined by coating with whitewash or washable distemper, and drying. Plug bolt holes, water jacket apertures, etc., with asbestos cement or similar filler, and dry.

## (2) TINNING

Use pure tin for tinning steel and cast iron surfaces; use 50% tin, 50% lead solder for tinning bronze, gunmetal or brass surfaces.

Flux surfaces to be tinned by swabbing with 'killed spirit' (saturated solution of zinc in concentrated commercial hydrochloric acid, with addition of about 5% free acid), or suitable proprietary flux.

Tinning cast iron presents particular difficulty due to the presence of graphite and, in the case of used bearings, absorption of oil. It may be necessary to burn off the oil, scratch brush, and flux repeatedly, to tin satisfactorily. Modern methods of manufacture embodying molten salt bath treatment to eliminate surface graphite enable good tinning to be achieved, and such bearings may be retinned several times without difficulty.

### Tin bath

(i) Where size of bearing permits, a bath of pure tin held at a temperature of 280°–300°C or of solder at 270°–300°C should be used.
(ii) Flux and skim surface of tinning metal and immerse bearing only long enough to attain temperature of bath. Prolonged immersion will impair bond strength of lining and cause contamination of bath, especially with copper base alloy housings or shells.
(iii) Flux and skim surface of bath to remove dross, etc., before removing bearing.
(iv) Examine tinned bearing surface. Wire brush any areas which have not tinned completely, reflux and re-immerse.

## Stick tinning

(i) If bearing is too large, or tin bath is not available, the bearing or shell should be heated by blowpipes or over a gas flame as uniformly as possible.

(ii) A stick of pure tin, or of 50/50 solder is dipped in flux and applied to the surface to be lined. The tin or solder should melt readily, but excessively high shell or bearing temperatures should be avoided, as this will cause oxidation and discoloration of the tinned surfaces, and impairment of bond.

(iii) If any areas have not tinned completely, reheat locally, rub areas with sal-ammoniac (ammonium chloride) powder, reflux with killed spirit, and retin.

## (3) LINING METHODS

### (a) Static lining

#### (i) Direct lined bearings

The lining set-up depends upon the type of bearing. Massive housings may have to be relined *in situ*, after preheating and tinning as described in sections (1) and (2). In some cases the actual journal is used as the mandrel (see Figures 19.1 and 19.2).

Journal or mandrel should be given a coating of graphite to prevent adhesion of the whitemetal, and should be preheated before assembly.

Sealing is effected by asbestos cement or similar sealing compounds.

#### (ii) Lined shells

The size and thickness of shell will determine the type of lining fixture used. A typical fixture, comprising face plate and mandrel, with clamps to hold shell, is shown in Figure 19.5 while Figure 19.6 shows the pouring operation.

Figure 19.1 Location of mandrel in end face of direct lined housing

Figure 19.3 Direct lined housing. Pouring of white-metal

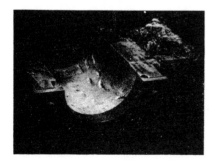

Figure 19.2 Outside register plate, and inside plate machined to form radius

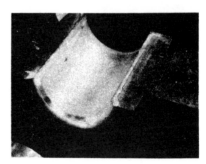

Figure 19.4 Direct lined housing, as lined

### (b) Centrifugal lining

This method is to be preferred if size and shape of bearing are suitable, and if economic quantities require relining.

### (i) Centrifugal lining equipment

For small bearings a lathe bed may be adapted if suitable speed control is provided. For larger bearings, or if production quantities merit, special machines with variable speed control and cooling facilities, are built by specialists in the manufacture or repair of bearings.

### (ii) Speed and temperature control

Rotational speed and pouring temperature must be related to bearing bore diameter, to minimise segregation and eliminate shrinkage porosity.

Rotational speed must be determined by experiment on the actual equipment used. It should be sufficient to prevent 'raining' (i.e. dropping) of the molten metal during rotation, but not excessive, as this increases segregation. Pouring temperatures are dealt with in a subsequent section.

### (iii) Cooling facilities

Water or air–water sprays must be provided to effect directional cooling from the outside as soon as pouring is complete.

*Figure 19.5 Lining fixture for relining of shell type bearing*

*Figure 19.6 Pouring operation in relining of shell type bearing*

### (iv) Control of volume of metal poured

This is related to size of bearing, and may vary from a few grams for small bearings to many kilograms for large bearings.

The quantity of metal poured should be such that the bore will clean up satisfactorily, without leaving dross or surface porosity after final machining.

Excessively thick metal wastes fuel for melting, and increases segregation.

### (v) Advantages

Excellent bonding of whitemetal to shell or housing.
Freedom from porosity and dross.
Economy in quantity of metal poured
Directional cooling.
Control of metal structure.

### (vi) Precautions

High degree of metallurgical control of pouring temperatures and shell temperatures required.

Close control of rotational speed essential to minimise segregation.

Measurement or control of quantity of metal poured necessary.

Control of timing and method of cooling important.

*Figure 19.7 Purpose-build centrifugal lining machine for large bearings*

*Figure 19.8 Assembling a stem tube bush 680 mm bore by 2150 mm long into a centrifugal lining machine*

## (4) POURING TEMPERATURES

### (a) Objective

In general the minimum pouring temperature should be not less than about 80°C above the liquidus temperature of the whitemetal, i.e. that temperature at which the whitemetal becomes completely molten, but small and thin 'as cast' linings may require higher pouring temperatures than thick linings in massive direct lined housings or large and thick bearing shells.

The objective is to pour at the minimum temperature consistent with adequate 'feeding' of the lining, in order to minimise shrinkage porosity and segregation during the long freezing range characteristic of many white-metals. Table 19.2 gives the freezing range (liquidus and

commence at the bottom and proceed gradually upwards, and the progress of solidification may be felt by the puddler. When freezing has nearly reached the top of the assembly, fresh molten metal should be added to compensate for thermal contraction during solidification, and any leakage which may have occurred from the assembly.

### (d) Cooling

Careful cooling from the back and bottom of the shell or housing, by means of air–water spray or the application of damp cloths, promotes directional solidification, minimises shrinkage porosity, and improves adhesion.

**Table 19.2 Whitemetals, solidification range and pouring temperatures**

| Specification | Nominal composition % | | | | | Solidus temp °C | Liquidus temp °C | Min. pouring temp °C |
|---|---|---|---|---|---|---|---|---|
| | Antimony | Copper | Other | Tin | Lead | | | |
| ISO 4381 | 12 | 6 | 0 | Remainder | 2 | 183 | 400 | 480 |
| Tin base | 7 | 3 | 0 | Remainder | 0.4 | 233 | 360 | 440 |
| Alloys | 7 | 3 | 1.0 Cd | Remainder | 0.4 | 233 | 360 | 440 |
| ISO 4381 | 14 | 0.7 | 1.0 As | 1.3 | Remainder | 240 | 350 | 450 |
| Lead base | 15 | 0.7 | 0.7 Cd 0.6 As | 10 | Remainder | 240 | 380 | 480 |
| Alloys | 14 | 1.1 | 0.5 Cd 0.6 As | 9 | Remainder | 240 | 400 | 480 |
| | 10 | 0.7 | 0.25 As | 6 | Remainder | 240 | 380 | 480 |

solidus temperatures) and recommended minimum pouring temperatures of a selection of typical tin-base and lead-base whitemetals. However, the recommendations of manufacturers of proprietary brands of whitemetal should be followed.

### (b) Pouring

The whitemetal heated to the recommended pouring temperature in the whitemetal bath, should be thoroughly mixed by stirring, without undue agitation. The surface should be fluxed and cleared of dross immediately before ladling or tapping. Pouring should be carried out as soon as possible after assembly of the preheated shell and jig.

### (c) Puddling

In the case of large statically lined bearings or housings, puddling of the molten metal with an iron rod to assist the escape of entrapped air, and to prevent the formation of contraction cavities, may be necessary. Puddling must be carried out with great care, to avoid disturbance of the structure of the freezing whitemetal. Freezing should

## (5) BOND TESTING

The quality of the bond between lining and shell or housing is of paramount importance in bearing performance. Non destructive methods of bond testing include:

### (a) Ringing test

This is particularly applicable to insert or shell bearings. The shell is struck by a small hammer and should give a clear ringing sound if the adhesion of the lining is good. A 'cracked' note indicates poor bonding.

### (b) Oil test

The bearing is immersed in oil, and on removal is wiped clean. The lining is then pressed by hand on to the shell or housing adjacent to the joint faces or split of the bearing. If oil exudes from the bond line, the bonding is imperfect.

### (c) Ultrasonic test

This requires specialised equipment. A probe is held against the lined surface of the bearing, and the echo pattern resulting from ultrasonic vibration of the probe is observed on a cathode ray tube. If the bond is satisfactory the echo occurs from the back of the shell or housing, and its position is noted on the C.R.T. If the bond is imperfect, i.e. discontinuous, the echo occurs at the interface between lining and backing, and the different position on the C.R.T. is clearly observable. This is a very searching method on linings of appropriate thickness, and will detect small local areas of poor bonding. However, training of the operator in the use of the equipment, and advice regarding suitable bearing sizes and lining thicknesses, must be obtained from the equipment manufacturers.

This method of test which is applicable to steel backed bearings is described in ISO 4386-1 (BS 7585 Pt 1). It is not very suitable for cast iron backed bearings because the cast iron dissipates the signal rather than reflecting it. For this material it is better to use a gamma ray source calibrated by the use of step wedges.

### (d) Galvanometer method

An electric current is passed through the lining by probes pressed against the lining bore, and the resistance between intermediate probes is measured on an ohm-meter. Discontinuities at the bond line cause a change of resistance. Again, specialised equipment and operator training and advice are required, but the method is searching and rapid within the scope laid down by the equipment manufacturers.

### (6) LOCAL REPAIR BY PATCHING OR SPRAYING

In the case of large bearings, localised repair of small areas of whitemetal, which have cracked or broken out, may be carried out by patching using stick whitemetal and a blowpipe, or by spraying whitemetal into the cavity and remelting with a blowpipe. In both cases great care must be taken to avoid disruption of the bond in the vicinity of the affected area, while ensuring that fusion of the deposited metal to the adjacent lining is achieved.

The surface to be repaired should be fluxed as described in section (2) prior to deposition of the patching metal. Entrapment of flux must be avoided.

The whitemetal used for patching should, if possible, be of the same composition as the original lining.

Patching of areas situated in the positions of peak loadings of heavy duty bearings, such as main propulsion diesel engine big-end bearings, is not recommended. For such cases complete relining by one of the methods described previously is to be preferred.

## THE PRINCIPLE OF REPLACEMENT BEARING SHELLS

Replacement bearing shells, usually steel-backed, and lined with whitemetal (tin or lead-base), copper lead, lead bronze, or aluminium alloy, are precision components, finish machined on the backs and joint faces to close tolerances such that they may be fitted directly into appropriate housings machined to specified dimensions.

The bores of the shells may also be finish machined, in which case they are called 'prefinished bearings' ready for assembly with shafts or journals of specified dimensions to provide the appropriate running clearance for the given application.

In cases where it is desired to bore *in situ*, to compensate for misalignment or housing distortion, the shells may be provided with a boring allowance and are then known as 'reboreable' liners or shells.

The advantages of replacement bearing shells may be summarised as follows:

(1) Elimination of hand fitting during assembly with consequent labour saving, and greater precision of bearing contour.
(2) Close control of interference fit and running clearance.
(3) Easy replacement.
(4) Elimination of necessity for provision of relining and machining facilities.
(5) Spares may be carried, with saving of bulk and weight.
(6) Lower ultimate cost than that of direct lined housings or rods.

### Special Note

'Prefinished' bearing shells must not be rebored *in situ* unless specifically stated in the maker's catalogue, as many modern bearings have very thin linings to enhance load carrying capacity, or may be of the overlay plated type. In the first case reboring could result in complete removal of the lining, while reboring of overlay-plated bearings would remove the overlay and change the characteristics of the bearing.

Linings are attached to their shoes by riveting or bonding, or by using metal-backed segments which can be bolted or locked on to the shoes. Riveting is normally used on clutch facings and is still widely used on car drum brake linings and on some industrial disc brake pads. Bonding is used on automotive disc brake pads, on lined drum shoes in passenger car sizes and also on light industrial equipment.

For larger assemblies it is more economical to use bolted-on or locked-on segments and these are widely used on heavy industrial equipment. Some guidance on the selection of the most appropriate method, and of the precautions to be taken during relining, are given in the following tables.

### Table 20.1 Ways of attaching friction material

| | Riveting | Bonding | Bolted-on-segments | Locked-on-segments |
|---|---|---|---|---|
| *Shoe relining* | De-riveting old linings and riveting on new linings can be done on site. General guidance is given in BS 3575 (1981) SAE J660 | Shoes must be returned to factory. Cannot be done on site | Can be relined on site without dismantling brake assembly. Bolts have to be removed | Can be relined on site without dismantling brake assembly by slackening off bolts |
| *Plate clutch relining* | As above | As above | Not applicable | Not applicable |
| *Use of replacement shoes already lined* | Quick. Old shoes returned in part exchange | Quick. Old shoes returned in part exchange | Quick | Quick |
| *Use of replacement clutch plates already lined* | As above | As above | Not applicable | Not applicable |
| *Friction surface* | Reduced by rivet holes | Complete unbroken surface, giving full friction lining area | As for bonding apart from bolt slots in side of lining | As for bonding |
| *Life* | Amount of wear governed by depth of rivet head from working surface. If linings are worn to less than 0.8 mm (0.031 in) above rivets they should be replaced | Not affected by rivets or rivet holes. Can be worn right down. If linings are worn to less than 1.6 mm (0.062 in) above shoe they should be replaced | Governed by thickness of tie plates. Advantage over rivets, or countersunk screws. If linings are worn to less than 0.8 mm (0.031 in) above tie plates they should be replaced | Governed by depth of keeper plates. Comparable with use of rivets or countersunk screws. If linings are worn to less than 0.8 mm (0.031 in) above countersunk screws they should be replaced |

**Table 20.1** (continued)

|  | Riveting | Bonding | Bolted-on segments | Locked-on segments |
|---|---|---|---|---|
| *Spares* | Good. Linings drilled or undrilled can be supplied ex-stock together with rivets. Small space required for stocks | Bulky. Complete shoe or plate with lining attached required. Where large metal shoes or plates are involved there is a high cost outlay and extra storage space | Good. Only linings with tie plates bonded into them required | Good. Only linings required suitably grooved |
| *Limitations* | Less suitable for low-speed, high-torque applications | Less suitable for high ambient temperatures, corrosive atmospheres, or where bonding to alloy with copper content of over 0.4% | 12 mm ($\frac{1}{2}$ in) thick or over, up to 610 mm (24 in) long | 12 mm ($\frac{1}{2}$ in) thick or over, up to 610 mm (24 in) long |
| *Suitable applications* | General automotive and industrial | Large production runs. Attachment of thin linings. Attachment to shoes where other methods are not practicable | For attachment to shoes which are not readily dismantled such as large winding engines and excavating machinery brakes, also high torque applications | As for bolted-on segments |

**Table 20.2  Practical techniques and precautions during relining**

|  | Riveting | Bonding | Bolted-on segments | Locked-on segments |
|---|---|---|---|---|
|  | | | | |
| *Removing lining or facings* | Best to strip old linings and facings by drilling out rivets, taking care to avoid damage to the rivet holes and shoe platform | Best done as a factory job | Slacken off brake adjustment. Slacken off nuts on main side. Remove nuts and bolts from outer side and slide linings across the slots in side of the lining | Slacken off brake adjustment. Slacken off bolts sufficiently to allow linings to slide along the keeper plates. If necessary tap the linings with a wooden drift to assist removal |
| *Replacing lining or facings* | Clean shoes and spinner plate, replace if distorted or damaged. If new linings or facings are drilled clamp to new shoe or pressure plate, insert rivets, clench lightly. Insert all rivets before securely fastening. If undrilled, clamp to shoe or spinner plate, locate in correct position and drill holes using the drilled metal part as a template. Counterbore on opposite side. Use same procedure as for drilled linings or drilled facings to complete the riveting. During relining particular attention should be given to rivet and hole size and also to the clench length | See above | Replace by the reverse procedure using the slots to locate the linings. Tighten up all bolts and readjust the brake | Replace by the reverse procedure. Afterwards readjust the brake |

### Table 20.2 (continued)

| | Riveting | Bonding | Bolted-on segments | Locked-on segments |
|---|---|---|---|---|
| Precautions | Use brass or brass-coated steel rivets to avoid corrosion problems. Copper rivets may be used for passenger car linings and also in light industrial applications. There must be good support for the rivet and the correct punch must be used. Allow one third the thickness of lining material under rivet head | Best done as a factory job | Best to use high-tensile steel socket-head type of bolt. Can be applied to all linings over 12 mm ($\frac{1}{2}$ in) thick | Avoid over-tightening the bolts so as not to distort the heads or crack the linings. Generally linings must be 12 mm ($\frac{1}{2}$ in) thick or over for rigid material and at least 25 mm (1 in) thick for flexible material |

Care must be taken during relining to avoid the lining becoming soaked or contaminated by oil or grease, as it will be necessary to replace the lining, or segment of lining, for if not its performance will be reduced throughout its life

### Table 20.3 Methods of working the lining and finishing the mating surfaces

| Material | Cutting | Drilling | Surface finishing | Handling |
|---|---|---|---|---|
| Woven materials | Hacksaw or bandsaw. Grinding is not recommended as it gives a scuffed surface and possible fire hazard | HSS tools are suitable. A burnishing tool is necessary to remove ragged edges | HSS tools are suitable for turning or boring of facings | Can be more easily bent to radius by heating to around 60°C. When machining and handling asbestos based materials, work must be carried out within the relevant asbestos dust regulations. |
| Moulded materials | Hacksaw, bandsaw or abrasive wheels | Tungsten carbide (WC) tipped tools are usually needed | Tungsten carbide (WC) tipped tools are usually needed for turning or boring of facings | Care must be taken to give adequate support during machining because of their brittle nature |
| Mating members | Fine/medium ground to 0.63–1.52 $\mu$m (25–60 $\mu$in) cla surface finish is best. Avoid chatter marks, keep drum ovality to within 0.127 mm (0.005 in) and discs parallel to within 0.076 mm (0.0003 in). Surface should be cleaned up if rust, heat damage or deep scoring is evident but shallow scoring can be tolerated. If possible the job should be done *in situ* or with discs or drums mounted on hubs or mandrels. The total amount removed by griding from the disc thickness or the drum bore diameter should not exceed: 1.27 mm (0.05 in) on passenger cars; 2.54 mm (0.1 in) on commercial vehicles. If these values are exceeded a replacement part should be fitted. When components have been ground, a thicker lining should be fitted to compensate for the loss of metal. With manual clutches the metal face can be skimmed by amounts up to 0.25 mm (0.01 in). For guidance on the reconditioning of vehicle disc and drum brakes, reference should be made to the vehicle manufacturer's handbook. | | | |

# Industrial flooring materials

## Factors to consider in the selection of a suitable flooring material

| Factor | Remarks |
| --- | --- |
| Resistance to abrasion | This is usually the most important property of a flooring material because in many cases it determines the effective life of the surface. Very hard materials which resist abrasion may, however, have low impact resistance |
| Resistance to impact | In heavy engineering workshops this is often the determining factor in the choice of flooring material |
| Resistance to chemicals and solvents | In certain industrial environments where particular chemicals are likely to be spilled on the floor, the floor surfacing must not be attacked or dissolved |
| Resistance to indentation | Any permanent indentation by shoe heels or temporarily positioned equipment is unsightly, makes cleaning difficult and, in severe cases, can cause accidents. |
| Slipperiness | Slipperiness depends not only on the floor surfacing material, but on its environment. Cleanliness is important. Any adjacent floors which are wax polished, can result in wax layers being transferred to an otherwise non slip surface by foot traffic. Adjacent floors with different degrees of slipperiness can cause accidents to unaccustomed users |
| Other safety aspects | Potholing, cracking and lifting can occur in badly laid floors. In high fire risk areas, floors which do not generate static electricity are required. Non absorbent floors are normally necessary in sterile areas |
| Ease of cleaning | This is a key factor in total flooring cost, and in maintaining the required properties of the floor |
| Comfort | In light engineering workshops, laboratories and offices, comfort can usually be taken more into account without sacrificing the performance of the floor from other aspects |
| Initial laying | The standard of workmanship and the familiarity of trained operators with the laying process can have a major effect on floor performance. Faults in foundations, or in a sub floor can result in faults in the surface |
| Subsequent repair | This must be considered when selecting a floor material, particularly in applications where damage is inevitable. Small units like tiles can usually be repaired quickly. Asphalt floors can be used as soon as they are cool, but need space for heating equipment and specialised labour. Cement and concrete can be repaired by local labour, but production time is lost while waiting for hardening and drying |
| Cost | The initial relative cost of different materials should be compared with their probable life. British Standards and Codes of Practice describe non proprietary materials. Manufacturers of proprietary materials often have independent test data available |

## Comparative properties of some common floor finishes

| Type of finish | Wear resistance | | | | | Resistance to | | |
| --- | --- | --- | --- | --- | --- | --- | --- | --- |
| | Abrasion | Impact | Resistance to indentation | Slipperiness | Ease of cleaning | Acid | Alkalis | Sulphates |
| Portland Cement concrete | G—P | G—P | VG | G—F | F | VP | G | VP |
| Portland cement precast | G | G—F | VG | G—F | F | VP | G | P |
| High alumina cement concrete | G—P | G—P | VG | G—F | F | VP | P | G |
| Granolithic concrete | VG—F | G | VG | G—F | G | VP | P | P |
| Mastic asphalt | G—F | G—F | F—P | G—F | G—F | VG—F | G—F | VG |
| Cement bitumen | G—F | G—F | F—P | G—F | G—F | P | G | P |
| Pitch mastic | G—F | G—F | F—P | G—F | G—F | G—F | G—F | VG |
| Steel or cast iron tile | VG | VG | VG | F | G—F | VP | F | F—P |
| Steel anchor plates in PCC | VG | VG[1] | VG | G—F | F | VP | G | VP |
| Steel grid in PCC | VG | G[2] | VG | G[2] | F—P | VP | G | VP |
| Steel grid in mastic asphalt | G | G—F | G | G | F—P | P | G | G |
| Rubber sheet | G | VG | VG | G—F[3] | G | F—P | F—P | VG |
| Linoleum sheet or tile | G—F | F | F | G | G | P | P | G |
| PVC sheet and tile | G | G—F | G | G | G | G | G | G |
| Magnesium oxy chloride | G—F | F | F | F | G | P | F | F |
| Terrazzo | G | G—F | VG | G[4] | G | P | G—F | G—F |
| Thermo plastic tile | F | F | F | G | G[5] | F | F | G |
| Timber softwood board | F | G | F | G | G | F—P | F—P | F—P |
| Timber softwood block | F | G | F | G | G | F—P | F—P | F—P |
| Timber hardwood strip | G | G | G | G | G | F | F | F |
| Timber hardwood block | G | G | G | G | G | F | F | F |
| Wood chip board and block | F | F | VG | G | G | F | F | F |
| Clay tile and bricks | G | G—F | VG | G—F[4] | G | G—F | G—F | VG—G |
| Cement PVA emulsion | G—F | G—F | F | G | G—F | P | G | F |
| Cement rubber latex | G—F | G—F | F | G | G—F | P | G | F |
| Composition blocks | G | G | VG | F | F | F | F | F |
| Concrete tiles | G | G—F | VG | G—F | F | VP | G | P |
| Cork carpet | G | G | P | VG | F | G | G | G |
| Cork tiles | G | G | P | VG | F | G | G | G |
| Granite slab | VG | VG | VG | G | G | G | G | VG |
| Sand stone slab | G—F | G—F | VG | G | G—F | G—P | G—P | G—F |
| Sawdust cement | F | F | F | G | F | P | P | VP |

VG = Very good    G = Good    F = Fair    P = Poor    VP = Very poor

[1] Particularly suitable for heavy engineering workshops
[2] The grid size should be chosen to expose sufficient concrete for non slip purposes, but in small enough areas to reduce damage by impact
[3] Rubber can be slippery when wet, particularly with rubber soled shoes
[4] Clay tiles and terrazzo become slippery when polished or oiled
[5] Thermoplastic tiles require a special type of polish

# Index

d on or before